THE ROYAL HORTICULTURAL SOCIETY
PRACTICAL GUIDES

GROWING FROM SEED

THE ROYAL HORTICULTURAL SOCIETY
PRACTICAL GUIDES

GROWING FROM SEED

ALAN TOOGOOD

A Dorling Kindersley Book

LONDON, NEW YORK, MUNICH,
MELBOURNE, DELHI

PROJECT EDITOR Candida Frith-Macdonald
ART EDITOR Martin Hendry

SERIES ART EDITOR Ursula Dawson

MANAGING EDITOR Anna Kruger
MANAGING ART EDITOR Lee Griffiths

DTP DESIGNER Louise Waller

PRODUCTION MANAGER Mandy Inness

First published in Great Britain in 2002
by Dorling Kindersley Limited,
80 Strand, London WC2R 0RL

A Penguin Company

A CIP catalogue for this book is available from the British Library.
ISBN 0 7513 3722 6

Reproduced by Colourscan, Singapore
Printed and bound by Star Standard Industries, Singapore

See our complete catalogue at
www.dk.com

CONTENTS

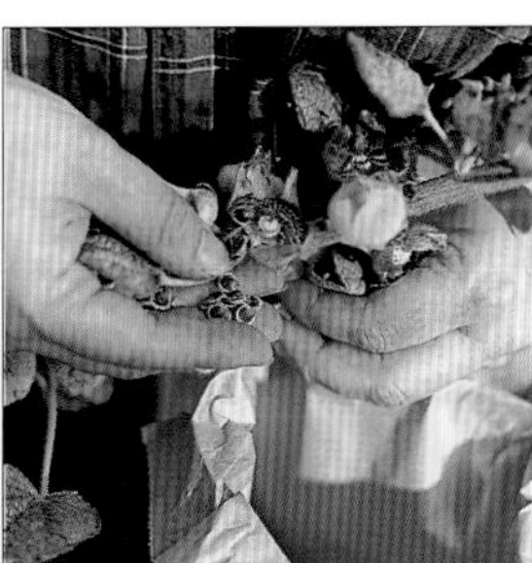

PLANTS GROWN FROM SEED

WHY GROW PLANTS FROM SEED?

THERE IS UNDOUBTEDLY SOMETHING dramatic about sowing apparently lifeless seeds and watching new, green plants emerge and grow. Perhaps because of this there is sometimes a certain mystique associated with raising plants from seed, as if it is particularly difficult or complicated. Although some seeds can be tricky, many garden plants reproduce easily by seed, and the advantages of raising your own plants are undeniable.

BENEFITS OF GROWING FROM SEED

Perhaps the first reason why gardeners grow from seed is cost. Seed from your own plants can be saved and sown, and bought seed costs less than bought plants. If you need large numbers of plants, for example for bedding, then growing your own from seed is far more economical than buying plants raised by someone else.

Seeds are also a good way to obtain new plants, perhaps those you have admired in the gardens of friends or neighbours, or even abroad; subject to certain regulations, it is far easier to bring home seeds than live plant material. Rare or unusual plants may be almost impossible to buy as growing specimens, and so raising from seed is the only practical way to obtain them.

ANNUAL RENEWAL *Annual plants such as this pink cosmos need to be renewed each year. Collecting seed from the plant and sowing it under ideal conditions will result in a large quantity of good-quality plants at much-reduced cost.*

SOW AND REAP *Crops such as these cabbages must be sown anew each year.*

SEEDS IN NATURE AND IN HISTORY

PLANTS HAVE EVOLVED a fascinating array of reproductive strategies in order to survive, increase, and colonize new ground. They have adapted to a wide range of adverse habitats, such as deserts, high altitudes where winds damage foliage and discourage pollinating insects, and even water, where problems are completely different. Since the dawn of civilization, farmers and gardeners have used observation of plant reproduction in the wild to grow plants in cultivation.

SEEDS IN THE WILD

Reproduction by seed remains the most important method of increase for many plants. Genetic material from a male and female of one species (preferably from different plants) unites in the seed or spore. The seed embryo forms a new plant that often looks the same as the parents, but has a different genetic make-up to either. This mixing of genes enables plants to adapt to changes or to colonize areas with different conditions. Another advantage is that embryos can lie dormant in conditions such as drought or a severe winter, germinating when more favourable conditions arise.

Reproduction by seed can give rise to new subspecies or varieties, which differ to some degree from the parents. This happens most in areas where plants become isolated, such as in mountain ranges or on islands. In contrast, where two species grow in the same area, they may cross-breed to produce natural hybrids.

Reproducing by seed allows plants to adapt and colonize new areas

In the wild, plants disperse hundreds or even millions of seeds to ensure that a few seedlings survive to maturity. In cultivation, a high yield of good-quality seedlings can be obtained by providing the most ideal environment possible (*see pp.34–35*).

IN THE WILD
The Cape area of South Africa is one of the world's richest areas for plants. It is especially diverse in species that reproduce by seed; the seeds lie dormant in the hot, dry season and germinate when the rains come. The seeds of some plants, such as these proteas, germinate after fire has cleared other growth, removing potential competition.

SEEDS IN CULTIVATION

Humans have also benefited from the genetic diversity of seeds, selecting forms that might not survive in the wild and developing valuable garden plants from them. Seeds offer the potential to introduce an exciting range of plants, with new forms of flower and leaf, or better hardiness, habit, adaptability for specific conditions, and resistance to pests and diseases.

The cultivation and propagation of plants began when humans began to live in settled communities. Ancient civilizations across the world grew a range of food crops, including cereals, from seeds. In ancient Greek and Roman times olives, date palms, cabbages, turnips, lettuces, and herbs were grown from seeds. The Greeks soaked seeds in milk or honey to speed up germination, and protected seeds with thin sheets of mica or a form of bell glass.

An explosion of plant-hunting took place in the western world in the 18th and 19th centuries. A wealth of new and exciting plants were discovered and traded between Europe and Asia, Africa, North America, Mexico, and South America, with many new introductions arriving as seeds.

Enthusiasm for these new plants and the desire to grow them led to the golden age of the glasshouse, which allowed the creative use of propagation methods and the refinement of techniques. The role of "propagator" became important for any garden of note; propagators were proud of their new knowledge and often guarded it jealously, to secure their reputations and future employment. This was probably the origin of the unnecessary mystique that often surrounds propagation even today.

▲ ROMAN SOWING
This Roman mosaic tile from the 3rd century shows a field being ploughed for sowing. The Romans were responsible for spreading many plants across Europe as their empire grew.

◀ ON TRIAL
Today seed companies have extensive trial grounds where new strains are grown in great numbers and evaluated before they are put on the market.

GROWING FOR IMMEDIATE IMPACT

THE GARDEN PLANTS MOST OFTEN grown from seed are annuals and biennials; indeed growing from seed is the only way these plants can be propagated. Because their entire life cycle from germination to setting their own seed is compressed into such a short time, they quickly grow to provide abundant colour. Some perennials are so quick to mature – and also so short-lived – that they are treated as annuals or biennials in the garden.

PLANTS FOR PLACES

Annuals and biennials can be grown in a huge range of situations. They may be used as part of a massed bedding scheme, grown in their own border, or included in mixed plantings. There are types to suit containers of all sorts, from bushy plants and climbers for pots to trailing plants ideal for hanging baskets, as well as tender plants for greenhouses or conservatories.

Because they die after setting seed, annuals and biennials rely entirely on their seeds for their continued survival. As a result of this, they are generally the easiest plants to raise from seed. Their seeds germinate readily and quickly, rarely having any dormant period or requiring any special treatment. Whether they are sown in containers and then planted out or are best sown in their flowering postion is dictated largely by the hardiness of the plants, the local climate, and how the established plants are to be displayed.

Annuals and biennials are generally the easiest plants to raise from seed

Young annuals develop very rapidly from seed, providing a fine display of colour soon after sowing – some annuals will flower within a few weeks. Biennials will not flower and die until their second year, and so need longer-term care than annuals:

MASSED BEDDING *This bed of massed annuals includes many reliable bedding favourites. Only the marigolds (tagetes) are true annuals; the begonias and dahlias are short-lived plants with perennials in their parentage that are bred to be treated as annuals.*

the seedlings must be grown on for a year, and are often raised in nursery beds for a season before being planted out into their final flowering position.

EASY EXPERIMENTS

Another consequence of the complete dependence of annuals and biennials on their seed is that many will produce prodigious quantities of seeds if they are suited to the climate. There is always some variation within species, and some species are more variable than others. Seeds saved from the garden will produce plants that may not always be completely true to type, but which are nonetheless usually pleasing and sometimes even surprising.

There is plenty of opportunity with these plants for selecting your own favourites; many of the commercial seed strains were developed by this kind of selection. If your garden has any trying conditions, such as exposure to wind, too much or too little rain, or a difficult soil, then seed saved from the plants that have grown best in your garden should have a higher chance of producing plants that will thrive than seed bought or brought from elsewhere.

▲ MIXED BORDER
This border contains both planned perennials, such as lilies and geraniums (cranesbill), and annual poppies and biennial foxgloves that have self-seeded throughout.

▼ SUMMER DISPLAY
Cascading bedding lobelias and a sprawling daisy-flowered Felicia bergeriana *combine to fill a window box within 12 weeks of sowing.*

Growing for the Table

Along with the excitement of raising a new plant, growing vegetables from seeds brings the added reward of an edible harvest, often within a few months. Culinary herbs offer a crop in even less time, providing aromatic flavours for dishes from summer salads to winter stews. Most crops are grown as annuals, raised from seed each year; you can buy in seed each year to try out different cultivars, or have the satisfaction of saving your own seeds.

Why Grow from Seed

Vegetable plants may be perennial, biennial, or annual, but as the whole plant is usually harvested, most are grown as annual crops. The main, and generally easy, method of propagation is from seeds, which may be sown in various ways depending on the crop and the climate. With some perennial vegetables, for example asparagus, only part of the plant is harvested, and these crops are left in the ground. Tuberous vegetables, particularly potatoes, are increased from "seed" tubers; these are not truly seeds and so are not included here.

Saving or Buying Seed

Always buy seeds that have been stored in cool conditions and are preserved in sealed packets. Purchased seeds are checked for viability, cleanliness, and purity before reaching the consumer, and are required to meet certain standards. As well as the traditional cultivars, many "F1" hybrids have now been developed by crossing two selected parents. These are more vigorous, produce larger crops, and may be of superior quality to open- or naturally pollinated cultivars. Research in recent years has enabled resistance to pests and diseases to be bred into many cultivars: some lettuces, for example, are resistant to

Resistance to pests and diseases has been bred into many cultivars

RAPID RETURNS *The traditional kitchen garden in full growth is an impressive sight, especially when you consider that all of these brassicas and other crops start from seed every year.*

◄ DESIGNER FOOD
Vegetables need not be hidden in some far corner of the garden; these red and green lettuces make fine summer bedding.

▼ SUMMER HERBS
Many herbs, such as basil, are easily raised from seed and can be kept in pots in or near the kitchen. Sow small amounts in succession for a continuing supply of fresh leaves.

lettuce root aphid and downy mildew; a number of cultivars of parsnip resist canker; and all modern tomato cultivars are resistant to tomato leaf mould.

You can also collect seeds from plants in your garden. Most vegetables are prevented from flowering to obtain a crop, but with some, such as leeks, it is worth allowing a few plants to run to seed to provide for next year's sowing. Some vegetables cross-pollinate very freely and produce variable offspring, but others will come fairly true to type from home-collected seeds. Seeds from F1 hybrids do not come true to type, but gardeners who are not concerned with uniformity can experiment with other types.

Sowing Vegetables

The traditional method of sowing vegetable seeds outdoors is in drills (*see pp.54–55*) in a vegetable plot, but they may also be sown in beds to avoid digging, in containers, or in informal patches in an ornamental kitchen garden. Some methods of seed sowing, such as fluid sowing and intercropping, are peculiar to the propagation of vegetables.

Vegetables are usually sown direct or quickly transplanted as seedlings into their permanent site. It is therefore particularly important to provide the optimum conditions for the best possible crop. This involves preparing the soil well, rotating crops to avoid build-up of pests and diseases, and sowing appropriate cultivars for the required harvest time. Vegetables are often classed as cool-, temperate- or warm-climate crops; the sowing times will vary depending on the climate.

Growing Specialist Plants

The more specialist types of plants include certain hardy perennials, ferns, alpines, bulbous plants, cacti, aquatics, and ornamental grasses. Many are easy to raise from seed given the right conditions, allowing gardeners to build up collections for a modest outlay. In some instances, raising from bought seed is the only way to obtain plants, particularly rare or unusual ones. Some plant collectors join specialist societies, many of which run seed-distribution schemes.

Reasons for Using Seeds

Perennials form a group of enormous value to the gardener, including not only the traditional border plants but also alpines, water garden plants, ferns, and ornamental grasses, including bamboos.

To give impact to plantings, perennials are often required in quantity. Many are easy to raise from seed, although new plants take longer to flower, and spores can similarly be used for ferns (*see pp.48–49*). Some more unusual plants may only be available or affordable as seeds, and perennials that are very slow to increase vegetatively, such as *Hepatica* and *Pulsatilla*, are all raised in large numbers from seed commercially. Growing from seed also offers the only way of raising monocarpic species, such as some verbascums, which die after flowering.

There are also many bulbous plants; these are most often bought as bulbs, but there are good reasons to raise them from seed. It is the recommended way of increasing many species and some tuberous plants that do not divide well. Woodland species, which dislike any drying out or root disturbance, are also best propagated from fresh seeds. Rare species are often only available as seeds. Young bulbous plants that you have raised yourself also settle well in the garden, which is not always the case with large, purchased plants.

Raising perennials or bulbous plants from seed requires patience

FERN GARDEN *Many ferns can be raised surprisingly easily from spores, although the more ornate cultivars will not generally come true to type and are better divided.*

◀ ALPINE BED
Although many of the more popular alpines are now widely available as container grown plants, some of the rarer types will only be available as seeds. Growing these unusual alpines from seed can be challenging, however, and is only for the committed.

▼ DRIFTS OF BULBS
Some bulbous plants are short-lived and replace themselves naturally by seed. Among them are crocus species that are excellent for naturalizing in grass.

Another good reason to propagate bulbous plants fom seed is plant health. Over time many of these plants lose vigour and can fall prey to disease, especially lilies and their close relations. These plants can always be renewed by seed-raised bulbs, as seeds are virus free, even if the parent is not.

Whatever your reasons for growing bulbous plants from seed, patience is required, because bulbous plants can take four years or more to reach flowering size.

Saving and Buying Seed

If buying seeds, particularly of rare plants, make sure that they come from a reputable source, such as specialist seed companies, or specialist or horticultural societies that operate seed distribution schemes.

Saving seeds from one's own plants is easily done: many perennials produce seeds readily, often in papery capsules or pods, and the seeds of most bulbous plants are large and easy to handle. Home-collected seeds do not always come true to type, but some cultivars do come reasonably true, including some delphiniums, lupins and oriental poppies (*Papaver orientale*). Seedlings of plants with coloured, marbled or variegated leaves, such as heuchera cultivars, vary in colour, so poor forms need to be rogued out at an early stage.

Growing for the Future

There are many slow-growing plants that can be grown from seed, always with the chance of creating something new. Trees, shrubs, and some climbers are a real investment in the future, but there are also herbaceous plants that may take years to flower, although the reward when it arrives is often spectacular. The sense of excitement as germination takes place and seedlings appear is the same however long it will take for the mature plant to develop.

Woody Plants from Seed

Trees tend to be expensive, because they are slow-growing compared with many herbaceous plants, so it is worth growing your own, especially if a number of plants are needed for hedges, woodland gardens, or screening. It also makes it possible to obtain more unusual species, replace aging trees, or to determine the size and shape of the tree. Trees naturally reproduce from seeds, so this is an easy and inexpensive way to raise species. Although only the dedicated will raise vast and slow-growing trees such as oaks from seed, faster-growing species such as eucalyptus and holly (*Ilex*) will provide useful plants in a few years. Hybrids and cultivars rarely come true to type, but natural seedling variation may always yield a new and interesting variety.

All the considerations that apply to trees also apply to both shrubs and woody climbers. Although some flower within a year or so from seed, others, such as rhododendrons, will take several years.

TEA TREE
Leptospermum lanigerum *is a shrub or small tree that is becoming more widely grown in Europe, and is easily raised from seed.*

Raising woody plants from seed is generally simple and inexpensive

Raising woody plants from seed is generally straightforward and inexpensive. Seedlings often establish well, and are unlikely to carry viruses from the parent plant. Seed-raised plants take two to five times as long to attain flowering size as those from cuttings, however, and may vary in appearance, hardiness, and growth. It is impossible to predict the sex of new plants, which is important for species in which only female plants have fruits, for example skimmias or hollies.

Success with these seeds depends as much on the treatment of seeds before sowing as on the sowing method. Many seeds

▲ SLOW TO MATURE
Although some of the plants in this border, such as the poppies, will flower the year after sowing, others, such as the alliums and the kniphofia, will take some years to do so.

▼ FAMILY TRAITS
Many species of sorbus will produce offspring identical to the parent plant from seed. They are easy to raise and flower in a few years.

germinate more successfully and faster if sown as soon as they ripen, but purchased seeds are adequate if stored correctly. Some seeds, especially those of the northern temperate regions, must be treated to break their natural dormancy before sowing.

Other Late Developers

It is not only woody plants that may keep you waiting for some years to see results. Although most perennials will flower two to three years after sowing, there are a few that will take even longer. Herbaceous peonies (*Paeonia*) are particularly slow, as are most hellebores and some kniphofias. Bulbous plants can also take several years to reach flowering size, most notably lilies, and many cacti are also slow to produce their first blooms; some cacti famously take many years, but most of those that are more commonly grown should flower within five years.

How Plants Produce Seed

The seed is the basic biological unit for the reproduction of both conifers and flowering plants. All of these plants produce pollen, which reaches and fertilizes the plants' ovaries by a variety of means. The fertilized ovaries then ripen and mature to produce seeds, on the scales of a cone in conifers, enclosed in a fruit in most other plants. Each seed combines male and female genes in a plant embryo that varies genetically from the parent plants.

The Structure of the Flower

Seed production begins with the flower: this contains male or female sex organs, or both. Most flowers have inner petals and outer sepals, in some plants collectively called tepals, and are diverse in shape and colour. Conifers have cone-like flowers, the males and females always separate.

The female reproductive part (carpel), which produces the seeds, is the ovary, connected by the style, a slender stalk, to the stigma, which receives pollen: there may be one or several of these, always at the centre or apex of the flower. Surrounding them in a bisexual flower are the stamens, the male part of the flower. Each stamen has a slender filament which supports the anther, where pollen is produced. Single-sex flowers have either stamens or carpels.

Pollination

To produce seeds, pollen from the anther is transferred to the stigma. Receiving pollen from another plant ensures healthy genetic variation in the seed, and most plants have systems to impede or prevent self-pollination. Some are monoecious, with single-sex flowers of both sexes on the same plant; sometimes they are on separate parts. Others, such as hollies (*Ilex*), and willows (*Salix*), are dioecious, with male and female flowers on different plants.

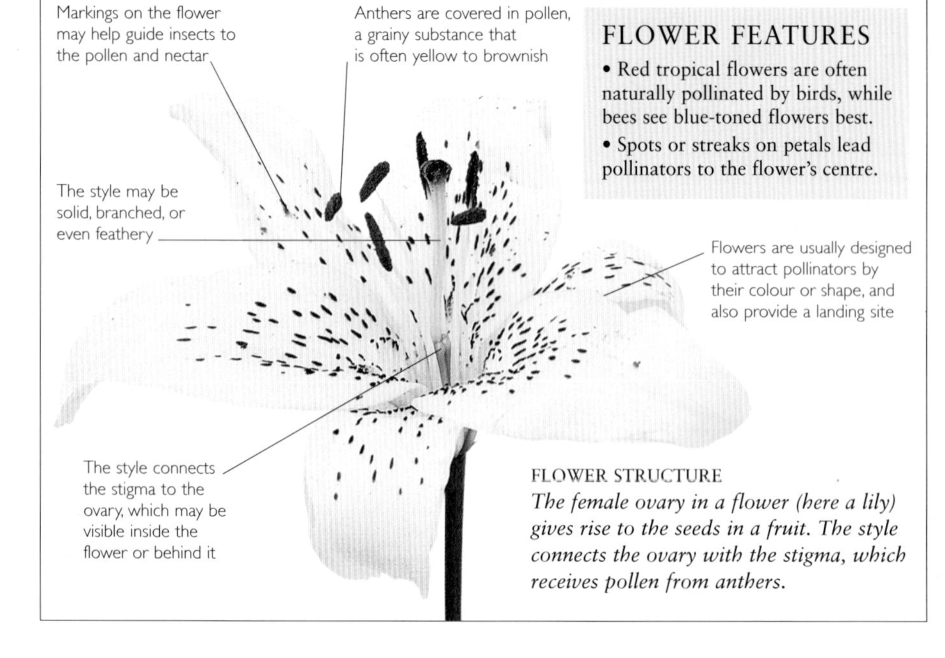

FLOWER FEATURES

- Red tropical flowers are often naturally pollinated by birds, while bees see blue-toned flowers best.
- Spots or streaks on petals lead pollinators to the flower's centre.

FLOWER STRUCTURE
The female ovary in a flower (here a lily) gives rise to the seeds in a fruit. The style connects the ovary with the stigma, which receives pollen from anthers.

INSECT POLLINATION
Many flowers are brightly coloured to attract pollinators such as this bee. Ripe pollen sticks to the bee and is carried to another flower.

Plants often exploit insects or animals to transfer pollen from one flower to another. The creatures are attracted by scent or by coloured or large petals and rewarded with nectar, protein-rich pollen, or fleshy petals. Bats, beetles, bees, butterflies, flies, small mammals, and moths all act as pollinators. Other plants use wind or water, so the flowers are often less showy because they need to offer no "bribe", but these methods are more wasteful and erratic. Conifers are generally wind-pollinated.

Fertilization of a Flower

If compatible, live pollen reaches a stigma that is receptive (usually it exudes a sugary solution and becomes sticky), the pollen will stick, germinate, and form a pollen tube. The tube grows down the style so that male cells can enter the ovary and fertilize the female egg cells (ovules). When this happens, seeds begin to form. Seeds usually consist of an embryo – a tiny plant with a shoot and a root, together with seed leaves, surrounded by a store of food, ready to become a new plant (*see pp.22–23*).

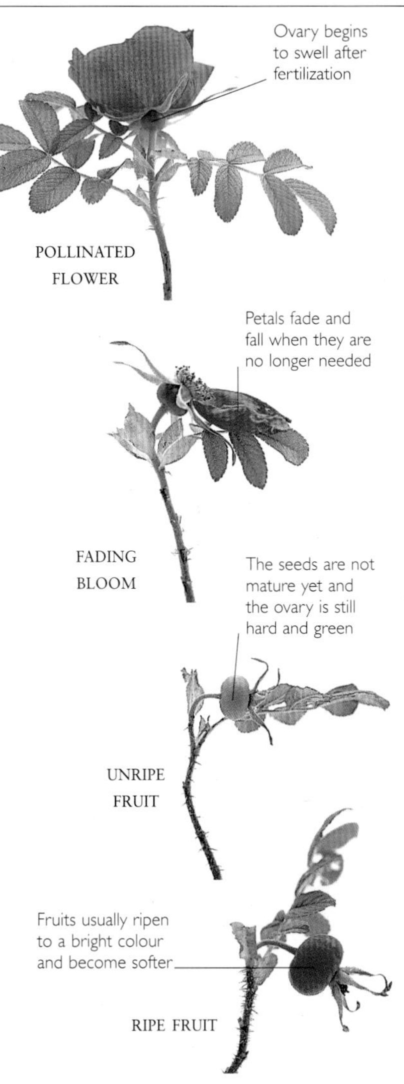

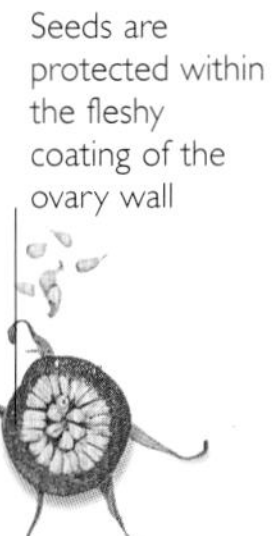

HOW SEEDS DEVELOP
Once the flower has been fertilized, the petals fade and fall and the ovary swells. The stigma and stamens wither and die. The fertilized egg cells (ovules) within the ovary develop seed coats, and the ovary wall forms a protective layer around the seeds.

SEEDS AND FRUITS

ONCE SEEDS HAVE MATURED, they must be dispersed; if they all germinated close to the parent plant, they would compete for water, nutrients, and light. Plants have developed various strategies to ensure that their seeds are dispersed far and wide, from having the seedpods open explosively and fling the seeds away or using structures that catch the wind or hook onto the coats of passing animals, to enclosing them in attractive fleshy fruits.

MECHANISMS OF SEED DISPERSAL

The fruits or pods that contain the seeds have adapted to different dispersal methods. Some fruits are very simple and look like a big seed, such as the oak acorn (*Quercus*), which has a thick shell to protect the true, thin-coated seed inside. Acorns are resistant to physical damage and can survive rolling around the ground and being buried by animals; when water eventually seeps though the wall they will germinate (*see pp.22–23*).

Some seed coats develop into papery capsules or pods, as are produced by delphiniums; the pod dries unevenly as it ripens, causing tension in the pod walls that

TYPES OF FRUIT

Fleshy fruits include drupes or stone fruits with single seeds, pomes with several seeds, compound fruits, and berries. Other fruits are dry pods or capsules and cones with the seeds held between the scales.

BLOWN AWAY
Some plants produce fluffy seedheads that contain small, light seeds with plumes, like this dandelion. The plumes enable the seeds to be carried over long distances on the wind. In this way, the plant can colonize very large areas.

eventually splits it open to release large numbers of seeds. The seeds either drop to the ground or are carried off on the wind.

Other seed pods, such as those of peas, burst explosively to expel the seeds over quite some distance. The successful weed, hairy bittercress, needs only to be touched or blown gently by the wind to cause its seed capsules to burst and eject seeds.

Seeds of some plants, for example grasses and other cereals, do not use these tactics. Their seeds germinate as soon as they ripen, while still on the parent plant, if conditions are suitably wet. The germinating seeds then fall into the moist soil and grow away immediately.

SPORE BEARING PLANTS

Plants such as mosses, liverworts, ferns, club mosses and horsetails reproduce by spores. A spore may look like a seed, but while a seed contains a new plant embryo within it (*see pp.18–19*), a spore develops male and female sex organs, independently from the plant that bore it. The sexual stage of reproduction takes place from the spore after it has left the parent plant, and can occur only in the presence of water (*see also pp.48–49*).

Seed Dispersal by Animals

Plants often have fleshy fruits to tempt animals to eat them. The seeds within pass unharmed through an animal's digestive system and are deposited in droppings (a ready-made seedbed) far away from the parent plant.

Many seeds and fruits have hooked spines that latch onto animal hair, feathers, or human clothing, some very tenaciously. These "burrs" may be transported over a great distance before being dislodged.

Wind and Water Dispersal

Many seeds are very small and carried by the wind. Minute seeds are produced in great numbers to compensate for the reduced likelihood of alighting on suitable soil. Other seeds have developed structures to keep them airborne: maples (*Acer*) have prominent, papery wings that spin like helicopter blades.

Plants adapted to growing in water or alongside watercourses produce seeds or fruits that are waterproof and buoyant. These are carried away by streams and rivers before they germinate: seeds of some coastal plants even survive ocean travel.

THE START OF A NEW PLANT

MUCH SEED IN THE WILD falls on inhospitable ground and fails to germinate. Therefore many plants produce huge quantities of seed to ensure that a few will develop into new plants. In cultivation, gardeners provide optimum conditions for germination and expect high percentages of seed to grow. Ideal germination conditions vary, depending on the plant, but in all cases moisture, air, sufficient warmth, and sometimes light are needed to trigger the process.

CONDITIONS FOR GERMINATION

Before a dried seed can begin to grow it must be rehydrated; water causes the seed coat to burst, and most seeds double in size before germinating. Massive amounts of oxygen are also needed to unlock the seed's energy reserves. If the soil or compost is frozen, compacted, waterlogged, or baked hard, oxygen will not reach the seed embryo.

Usually germination is prompted by temperatures typical of spring in the plant's natural habitat, giving the seedlings time to become established before the next winter. Suitable temperatures vary considerably: plants from tropical climates require much higher temperatures for germination than those from colder regions. Seeds from temperate climates usually germinate at 8–18°C (46–64°F), plants from warmer climates at 15–24°C (59–75°F). High temperatures can delay germination.

> Massive amounts of oxygen are needed for a seed to germinate

Some seeds need light for germination, especially very fine seeds that have little or no food stores to nourish the embryo, and nearly all seeds either die or become dormant if sown too deeply, because they cannot recognize when the surface light is sufficient for growth. As a rule of thumb, seeds are best covered to no more than their own depth.

Some seeds have more complex needs, and remain dormant until they are met. Fruit flesh may contain a chemical that inhibits germination and must be removed (*see p.29*), or the seed coat may need to be cut (*see p.33*). Other seeds require a winter's chill (*see p.31*). Some seeds from areas that have bush fires need smoke treatments, and are best bought ready-treated.

FROM THE ASHES
In areas prone to bush fires, seeds often lie dormant until fire destroys competing plants. The heat of fires splits open the fruits of some plants, like this banksia, and chemicals in the smoke trigger germination of the seeds.

How a Seed Germinates

There are two basic ways in which seeds germinate, and they make a difference to how quickly you can expect to see a seedling appear.

Plants such as the tomato and beech (*Fagus*) emerge by elevating the seed leaves above the surface at the same time as the root develops downwards. This is known as epigeal germination. This kind of germination will show a seedling quickly, but if the shoot tip is frosted or killed, no further growth is possible.

In plants such as the pea, oak (*Quercus*), and some bulbs, the seed leaves and food store remain in the soil with the root, and the growing shoot emerges only when the first true leaves form. This is known as hypogeal germination. The advantage is that if the seed is deep enough in the ground, it has a good chance of survival if the shoot tip is damaged, and can produce a secondary shoot or shoots. Hypogeal germination causes difficulty for gardeners, however, because it may be many months after germination before any sign of growth is visible on the surface.

With both kinds of germination, if levels of moisture, light, air, or warmth decline after the process starts, seeds quickly die.

OAK SEEDLING
Some seeds, such as the oak's acorns, only germinate after animals have damaged the shells, allowing water to reach the seed.

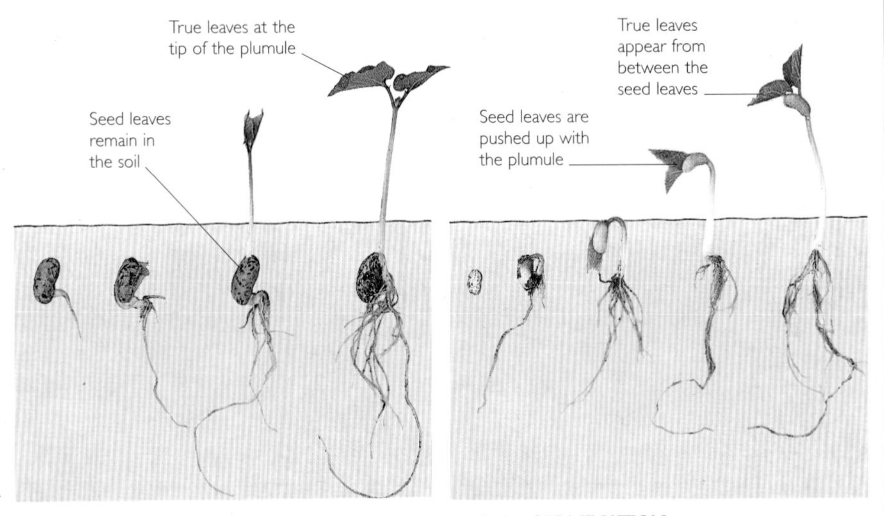

HYPOGEAL GERMINATION
Once the root emerges, the embryonic shoot (plumule) is pushed upwards, leaving the seed leaves behind in the soil. The plumule then emerges above the soil and produces its first true leaves.

EPIGEAL GERMINATION
The growth of the seed's root pushes the plumule and its protective seed leaves up out of the soil. The seed leaves are borne at the tip of the growing shoot until the first true leaves are produced.

Raising Plants from Seed

Before You Begin

Success in raising plants from seed usually depends on providing a supportive environment for the seeds and, later, for the new plants. Their needs are not generally difficult to meet, whether in the home, the greenhouse, or the garden. But it is important to know the basics: how to collect seeds, how to store them so they remain alive, the right way to sow them, and how to provide the optimum levels of light, warmth, and moisture for maximum success.

Collecting Seeds

Saving seeds from one's own garden plants is sometimes regarded as difficult or only for the truly dedicated gardener, but for generations most gardens were stocked this way, and it is both simple and rewarding. A large number of plants may be obtained at little cost, and seeds from the garden often produce plants that are better adapted to the local conditions: home collected vegetable seeds may be particularly adaptable. As a rule, collect seeds from species, not hybrids, as seedlings from hybrid plants can be extremely variable. Many vegetables, however, come reasonably true. With this proviso, seeds collected when ripe and stored and sown with care will yield good results.

GOOD TIMING *Picking seedheads when they are ripe is important: poppy seedheads picked when still green will not yield plants, but if picked when dry and about to open, a single seedhead can provide enough viable seed for dozens of new plants.*

◀ PROLIFIC PLANTS *Foxgloves self-seed well, but their seedlings may need thinning or moving.*

Easy Seedheads and Capsules

The best seed will come from a vigorous plant which seeds prolifically. Seeds should be collected at the point of ripeness, so check the ripening seedheads or capsules frequently; collecting them too early will mean that the seeds do not germinate at all. Always collect seedheads or capsules on a dry day, to avoid the risk of damp seeds rotting, and always label the seeds as soon as they are in a container, to avoid possible confusion later.

Collecting Seeds from Plants

Seedheads can ripen and release their seeds very quickly, so you will need to keep a careful watch and pick your time. Some plants have seed capsules that "explode", flinging their contents far from the parent plant; other plants have tiny seeds that are shed at the slightest disturbance, or plumed seeds that are carried off by the wind.

All of these seeds are best collected from the plant: bringing the seedheads indoors to remove the contents from them is likely to result only in a trail of seeds across the garden. Take paper bags, preferably already labelled, into the garden on a still, dry day and gather the seeds into them straight from the plant.

◀ OPENING SEEDHEADS
The opening seedheads of hollyhocks (Alcea) will shed their seeds when shaken, so do this over a paper bag.

▶ FLUFFY SEEDHEADS
Choose a dry day to gather plumed seeds. There is no need to detach the plumes before sowing.

SMALL SEEDS
Amaranthus seeds are ripe when the tassels begin to change colour. Gather the seeds by stroking or "milking" the tassels, while holding a tray beneath to catch the seeds.

COLLECTING SEEDS

Peppers (*Capsicum*) Dry the ripe fruits before extracting the seeds.

Clematis Fluffy seeds are easily carried off by the wind: collect straight from the plant.

Broom (*Cytisus*) Collect pods when ripe, but before they split open.

Foxgloves (*Digitalis*) Tiny, easily shed seeds: collect by shaking into a bag.

Eschscholzia Seed capsules explode when ripe: collect as they turn colour.

Impatiens Tie a bag over capsules when they change colour.

Poppies (*Papaver*) Gather capsules as they change colour and dry them.

Pine (*Pinus*) Collect ripe cones before the scales open and shed seed.

Extracting Fresh Seeds

Aim to collect ripe fruits as soon as the seedheads or capsules change colour, but before they open. Pick the just-ripened seedheads or capsules on a dry day, either singly or with their stalks. Many seeds can be removed very easily at this stage. Some seeds, particularly those of some summer-flowering perennials and bulbous plants, are short-lived and should be extracted and sown immediately (*see pp.30–31*).

SEEDS FROM FLOWERHEADS
Cut a sunflower that is about to go over and pick out the chaff from between the seeds. Flex the disc and stroke it firmly; the seeds will pop out.

SEEDS FROM GRASS SPIKES
Cut inflorescences that have fluffed up. The seeds should strip off each spike easily; if not, leave the stem in a cool, dry place for a few days.

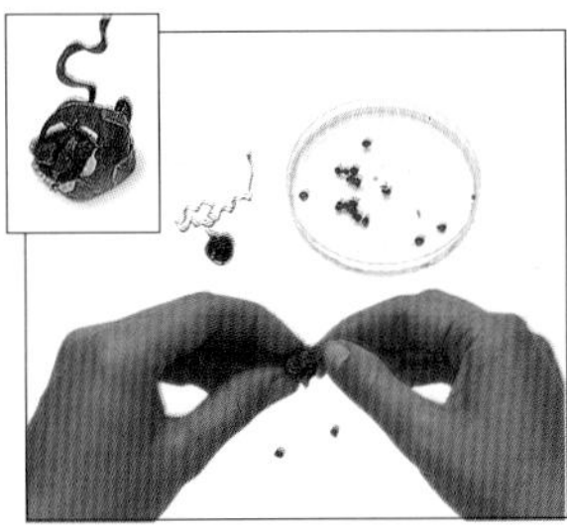

SEEDS FROM CAPSULES
Collect capsules, here from cyclamen, as they begin to open (see inset). Peel back the fleshy covering further and shake the seeds out.

Drying Pods and Capsules

Many pods and capsules can be dried to release the seeds. They should be stored in a warm, dry, and well-ventilated place. A warm windowsill or greenhouse bench is ideal for capsules in bowls or paper bags, while larger stems with many pods can be hung in a warm room or airing cupboard with a tray beneath them.

As capsules dry, holes or splits will open up to release the seeds

▲ LARGE CAPSULES
Dry capsules (here nigella) on blotting paper in a warm, sunny place. Shake tiny seeds out onto a piece of paper.

◀ BEAN PODS
In damp climates, gather pea and bean pods with their stems and roots intact. Hang them upside down in an airy, dry, dark, frost-free place until the pods are dry.

▶ DRYING IN BATCHES
Large quantities of seedheads or capsules can be dried in paper-lined boxes or trays.

DRYING TIPS

- Dry pods of peas, beans, or related plants in the dark.
- If plants have exploding capsules, dry them with a paper bag tied over them.
- Never pick unripe pods or seedheads and expect them to ripen while drying.

SPECIAL EXTRACTION TECHNIQUES

SEEDS ARE FORMED IN VARIOUS "containers" such as pods, capsules, fleshy fruits and berries, and woody cones. When the seeds are ripe they must be removed from these containers before they are sown or stored. The techniques of seed removal vary according to the type of container, and range from simply crushing dried capsules or pods to release the seeds, to pulping and soaking fleshy fruits in water to separate seeds from flesh.

SEPARATING SEEDS FROM CHAFF

Once seedheads or capsules have been dried (*see p.27*), they should open up to release the seeds. If they do not, crush them gently in your hand or a seive to break them open. The next stage is to separate the seeds from the broken fragments of their casing or the remains of the dead flowerhead, which is called the "chaff".

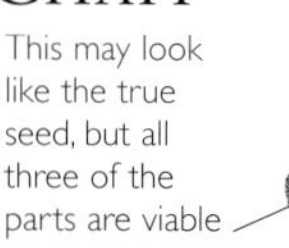

This may look like the true seed, but all three of the parts are viable

THREE-PART SEED
Calendula seeds often break into three parts when they are collected or stored. Each part can be sown, so do not discard them.

SIEVING AND BLOWING

- Where the seeds are large and relatively heavy, the simplest way of separating them is to place them in a bowl and gently blow the chaff off while stirring them.
- If you have a lot of seeds to process, it may be worth buying a set of stacking seed sieves, which have meshes of varying sizes.
- Do not use your kitchen sieves to clean seeds: some garden plants have toxic seeds.

SIMPLE SIEVE
Very fine seeds can be separated out with a tea strainer. Put the dried seedheads in the strainer over a sheet of clean paper, and gently break them up. The fine seeds will fall through the mesh, leaving the chaff in the strainer.

EXTRACTING SEEDS FROM DRIED BERRIES

Some berries can be dried for storage, with the seeds being extracted just prior to sowing. To dry berries, spread them out on a tray and keep it in a warm, dry, and well-ventilated place until they become hard and shrivelled. Placing them in an oven on a very low setting with the door slightly open is another alternative. Store the dry berries as you would dry seeds (*see p.30*).

CRUSHING BERRIES
Use a piece of wood to crush dried berries without damaging the seeds (here Actaea*). Separate the seeds from the remains of the dried fruit in a sieve, and sow.*

Obtaining Viable Seeds from Cones

Conifer fruits usually ripen in autumn, and often change colour in the process. In some junipers the cones change from green to blackish-purple or blue; in other conifers, such as cypress, immature cones may look very similar to ripe ones, but unripe seeds will not germinate. Cones may take one, two, or three summers to ripen, depending on the species. Cones that have fallen to the ground are likely to have shed their seed.

1 **Gather cones** that have changed colour but not opened up. Leave them in a labelled box in a warm place until they open.

VIABLE SEEDS

NON-VIABLE SEEDS

2 **Tip out seeds** onto a sheet of paper. Extract any that do not fall out with tweezers. Dark, plump seeds will be viable; discard thin, pale ones.

Extracting Seeds from Fleshy Fruits

Fruits usually soften and change colour as they ripen. Watch out for the turn: if you leave it too late, the soft, succulent fruit may be taken by birds. Collect fruits by hand-picking or by shaking the plant. Seeds can be removed from fruits or berries in many ways. Squeezing berries in muslin or gently mashing them through a sieve is the simplest way; you might also liquidize them and then wash off the pulp.

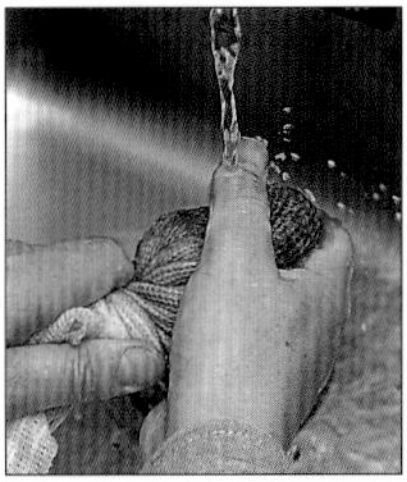

1 **For berries** with large seeds (here mahonia), place a handful in a cotton or muslin cloth, twist to secure, and hold under a cold running tap. Squeeze until no more juice runs out.

2 **Open out** the cloth and pick out the seeds from the fruit pulp. Leave the seeds to dry on kitchen or blotting paper in an airy place for a couple of days before sowing or storing them.

Extracting Seeds by Soaking

One way to extract seeds from small fleshy fruits or berries is to soak them in water. Pulp the fruits by gently crushing them with a block of wood, then soak the mixture in a container of warm water for up to four days to separate out the seeds. You may need to shake or stir the water occasionally to ensure separation. Viable seeds should sink to the bottom of the container and the flesh should float. Drain off the water and flesh, dry the clean seeds, and then sow or store them as appropriate.

WARM SOAK

Soak pulped fruits of both woody plants and perennials in water for several days to clean the flesh from the seeds. Change the water several times, as it should be warm or tepid for best results.

Caring for Seeds

Seeds are obviously "designed" to grow in the wild, but in cultivation they may need some help to germinate. Correct storage of collected seed goes a long way to ensuring good rates of germination. Seeds of many plants are given special treatments before sowing, such as chilling, also known as cold stratification. Purchased seed may also have been specially treated, either to ensure quick germination or to make it less prone to attacks by diseases.

Seed Longevity

Some seeds are extremely short-lived, and should be sown as soon as they are gathered. Many alpine seeds fall into this category, as do nuts and the seeds of water plants, both of which die if they dry out.

Other seeds remain viable for a year or two, or even longer if stored correctly. In general, seeds of annuals and biennials keep best; some peas and beans have germinated after thousands of years of storage.

THE LIFE OF SOME POPULAR SEEDS

SHORT LIFE	INTERMEDIATE	LONG LIFE
Anemone	Buddleja	Beans
Campanula	Broom (*Cytisus*)	Brassicas
Eucalyptus	Marigolds (*Tagetes*)	Cucumber
Ferns	Onions (*Allium*)	Grasses, ornamental
Foxglove (*Digitalis*)	Pot marigold (*Calendula*)	Lettuce
Maples (*Acer*)	*Prunus*	Mammillaria cacti
Poppies (*Papaver*)	Sunflowers (*Helianthus*)	Peas (*Lathyrus* and *Pisum*)
Primula	Sweet corn or maize (*Zea*)	Tomatoes
Rhododendron	Wallflower (*Erysimum*)	

Dry Storage

Most seeds should be stored dry to avoid the risk of fungal rot. Remove any damaged or shrivelled seeds before storing, as they are liable to be diseased. Put seeds in paper envelopes or bags and keep in a cool, dry place or in a sealed box in a refrigerator. A sachet of silica gel kept with the seeds will absorb any moisture.

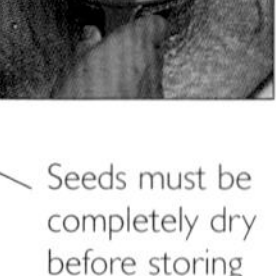

Seeds must be completely dry before storing

Individual labelled paper packets keep seeds separate

◄ BOXED UP
An airtight box will prevent the seeds from absorbing moisture from the air, and is vital in a humid refrigerator.

▲ CLEAN AND DRY
Rinse seeds from fleshy fruits, such as melons, thoroughly to remove any pulp. After cleaning, spread them out on kitchen paper and leave them in a warm, airy place to dry for 7–10 days.

Stratification

Some seeds, especially those of many woody plants, are prompted to germinate by temperature changes. Many need a winter's chilling before germinating in spring; mimic this by "cold stratification", storing seeds in a refrigerator before sowing. Even seeds that do not require chilling may germinate more quickly and evenly after cold storage. Some hard-coated seeds need "warm stratification" followed by cold stratification before sowing.

USING VERMICULITE
Put the seeds to be stored in a bag with moist vermiculite, label, and store them in a refrigerator for 4–12 weeks at 3–5°C (37–41°F).

USING BLOTTING PAPER
Some seeds, such as sorbus, germinate readily. Stratify on moist blotting paper, and check often. If any begin to germinate, sow them at once.

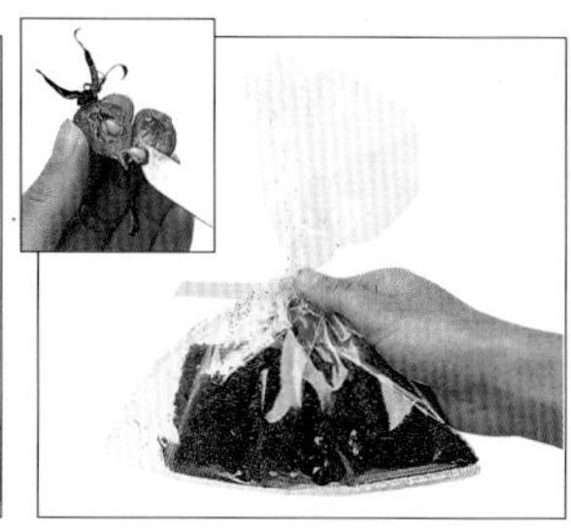

WARM STRATIFICATION
Put seeds in an equal volume of sand and leaf mould or soilless compost, and store them for 4–12 weeks at 20–25°C (68–77°F).

Purchasing Seed

As well as saving your own seed, you can choose from the wide array of seeds that are available commercially. These are sold in many forms, from the traditional packets (now usually sealed foil for freshness) to seeds that are treated, pelleted, embedded in tapes or sticks, or supplied in a kit with gel. Whatever form you choose, always check the date on the packaging to make sure that they not too old.

MODERN PACKAGES
Seeds now come in many forms to make sowing easiei and germination better.

TREATED AND UNTREATED SEEDS

Treatments are most often used on vegetable seeds. They may be "primed" to speed up germination, difficult seeds may be "chitted" or pregerminated, small seeds may be pelleted for easier handling, or those prone to mould may be coated or dusted with fungicide.

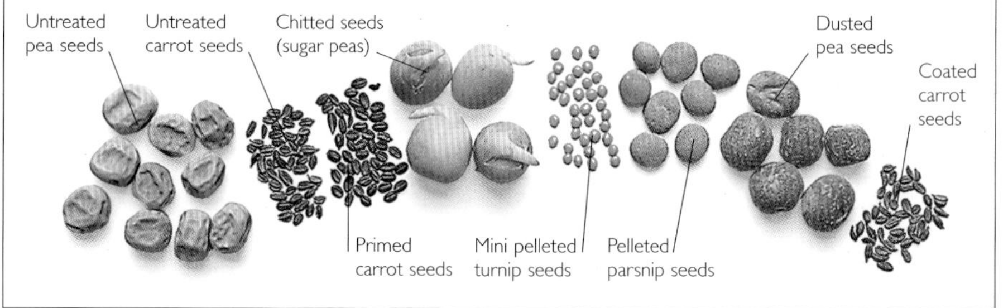

Priming Seeds

Some seeds have built-in dormancy to delay germination in the wild until the right conditions for seedling development arrive. They may have a tough seed coat that can only be softened by prolonged soaking, or contain chemicals that prevent germination until they have been washed off or broken down. There are several ways to break dormancy in seeds before sowing to obtain a good rate of germination.

Testing for Viability

The usual reason why seeds fail to grow is that dead seeds are sown. Seeds die for many reasons: they may not be fertilized or fully developed; they may be defective, especially if they are hybrid; they may have been stored too long, or be damaged by fungal or insect attack. After sowing, seeds may be killed by rot, rodents, or excessive cold or heat.

Dead seeds float because they are hollow

TESTING IN WATER
Put medium-sized or large seeds in a jar of water. Viable seeds sink; dead seeds float. Sow the viable seeds at once.

HOW VIABILITY VARIES

- Some seeds have very high rates of viability, and are hardly worth testing, such as beans, peas, sweet peas, grasses, and sunflowers.
- Fine, dustlike seed cannot be tested; most fresh seeds will be viable.
- Many seeds deteriorate in storage; always test seeds if they are not completely fresh
- A high proportion of conifer seeds are usually dead or infertile.

Washing and Soaking

Seeds are soaked to remove chemicals that inhibit germination. Some inhibitors can be removed simply by rinsing; others are more persistent, and the seeds must be soaked for at least two days to leach out the chemical. Seeds may also be soaked to soften a hard coating; this is a form of scarification. Whatever the reason for soaking seeds, do it immediately before sowing; soaked seeds will die if they are not sown at once.

◀ WASHING OFF INHIBITORS
The inhibitors on some seeds, such as these beetroot seed clusters, can be removed simply by rinsing them under cold running water just prior to sowing.

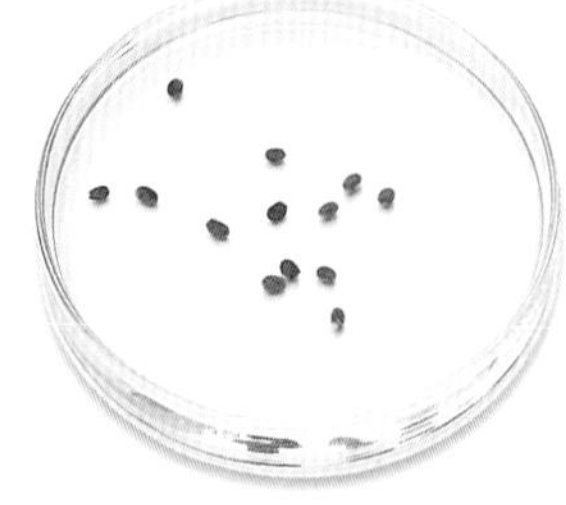

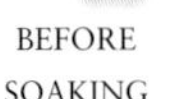

BEFORE SOAKING

AFTER SOAKING

▶ SOFTENING SEED COATS
Seeds with hard coats, such as these lupins, should be soaked for 24 hours in a saucer of cold water before sowing: they will swell up as their coats soften in the water.

SCARIFYING TECHNIQUES

Hard protective seed coats, found in shrubs, climbers, and perennials, are most common in the pea family. These seed coats must be softened or broken, or "scarified", in some way, so that moisture can enter. Filing seed coats is often advised, but this is painful and time-consuming with large numbers of seeds. A better way of scarifying larger seeds is to rub a batch with fine-grade sandpaper. The coats of seeds collected in cool, moist summers can be softened by soaking in cold water (see below left). If the seeds are large or come from plants grown in hot, dry conditions, use boiling water instead.

NICKING LARGE SEEDS
The hard coat of very large seeds (here of a peony) can be nicked using a sharp knife. Take care not to damage the "eye" of the seed or to cut too deeply (see inset); the aim is only to let water in.

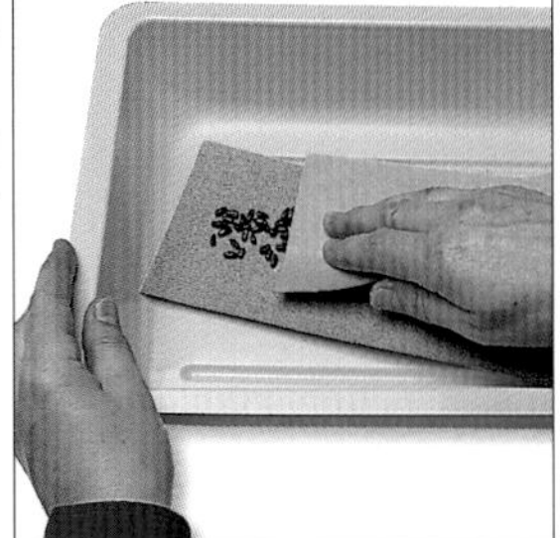

SANDING SMALL SEEDS
Place smaller, hard-coated seeds between two sheets of sandpaper in a seed tray and rub them to scratch and weaken their surfaces. This is also the best way to deal with large numbers of seeds.

USING HOT WATER
To soften the seedcoats of seeds produced in hot, dry conditions, place them in a bowl and pour boiling water over them. Allow the seeds to soak in the cooling water for 24 hours, then sow at once.

PREGERMINATION

Pregermination or "chitting" means making the seeds germinate before planting them. There are two advantages to this: firstly, you know that you are only planting viable seeds, so your success rate will be higher, and secondly the germination of some plants is speeded up, bringing results faster than would otherwise be expected. Pregermination is often used for vegetables, but it is also worth trying on other plants.

USING DAMP PAPER
Beans can be pregerminated on moist tissue at a minimum of 12°C (54°F). Sow the beans as soon as they produce shoots.

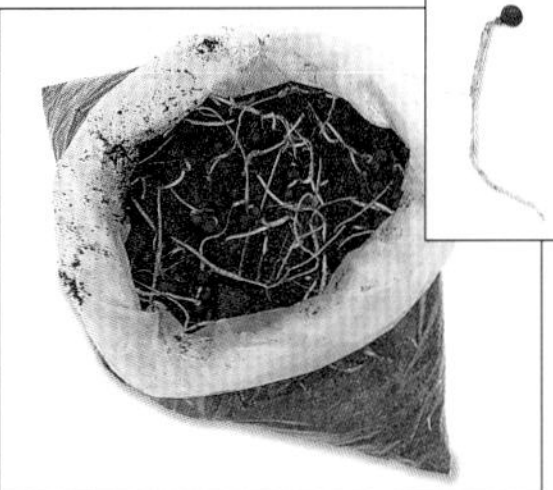

USING SOILLESS COMPOST
Pregerminate large seeds, especially those of tropical plants, in a bag of soilless compost in a warm place.

CHITTING SEEDS

- Large seeds are often not viable, so pregermination is usually worthwhile; this applies especially to palms and cycads.
- Seeds of cucumbers are difficult to germinate, and benefit from pregermination.
- Some root vegetables do best if chitted in a warm place and then fluid-sown outside in cooler conditions.

The Indoor Growing Environment

RAISING PLANTS INDOORS allows more control over environmental conditions and pests than sowing direct outdoors, and generally gives a higher success rate in raising healthy seedlings. It does not require a great deal of equipment: a surprisingly wide range of plants can be raised with nothing more than a few trays and pots on a sunny windowsill. Some items are essential, however, and there are others that will make the tasks involved quicker and easier.

Choosing the Right Container

A wide range of containers, including the traditional seed tray and pot, are now available. Seeds that make deep roots quickly can be sown into pots or modules. Plastic pots are more hygienic, lighter, and less expensive, but clay or terracotta pots provide better aeration and drainage. Seed trays are an efficient use of space for large numbers of seedlings, especially those that will be potted on or planted out quite young.

TRAYS AND POTS
The containers available for growing seedlings include modules, long used commercially, and biodegradable pots, as well as the more familiar trays and pots.

Pots for larger seeds

Seed tray for raising many seedlings

Biodegradable pot

Modules for growing individual seedlings

Useful Equipment

Plant labels are essential to keep track of sowings. These (or teaspoons) can be used quite effectively for pricking out seedlings, but dibbers and widgers designed for the purpose will be better. Pressers for firming compost – small for pots and large for seed trays – are also useful. A watering can with a fine rose is best for watering in seeds, and mist sprayers are useful for misting young plants that need a humid atmosphere.

COMPOST PRESSER
A small wooden presser with a handle is easily made; use a pot as a template.

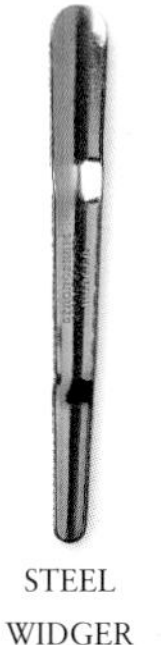

STEEL WIDGER

PLASTIC WIDGER

PLASTIC DIBBER

TOOLS AND TAGS
Dibbers are pencil-shaped tools, used to make planting holes; widgers allow lifting of seedlings with the minimum of disturbance. Labels are vital to keep track of sowings.

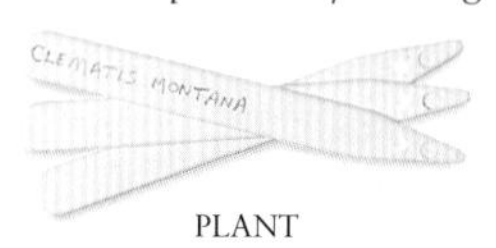

PLANT LABELS

The Importance of Hygiene

High standards of hygiene are essential to prevent pests and diseases from striking down your young plants. Keep your pots and other propagation equipment scrupulously clean by scrubbing them thoroughly after every use, and ideally sterilize them with boiling water or sterilizing solution immediately before making any new sowings into them. Wear gloves or wash your hands regularly while working, and keep work surfaces clean.

CLEANING POTS
Dirty containers can harbour diseases and minute pests. Wear protective gloves and using a stiff brush, scrub each pot thoroughly with dilute horticultural disinfectant. Rinse and allow to dry before use.

Under Cover

The essential needs of seeds and seedlings are light, warmth, and moist soil. These needs can be met by keeping containers on a bright windowsill or in a glassed-in porch or conservatory. The location provides light and warmth; moisture is maintained by covering the container with kitchen film, a sheet of glass or plastic, or a plastic bag. Propagators maintain moisture and can be heated; cold frames (*see p.46*) can be used for growing in; and with a greenhouse you can grow almost anything you want.

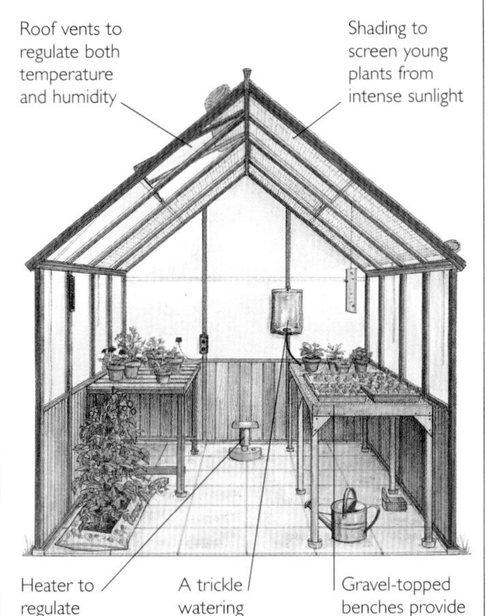

▲ THE GREENHOUSE
This is the most complete and controlled propagation environment, giving the chance to create several areas for different plants or varying uses.

USING PROPAGATORS

- Small windowsill propagators are better for indoors; large, heated propagators are useful in a greenhouse in cooler climates, creating higher temperatures.
- Adjustable vents in the lids allow moisture to escape and stop the atmosphere from becoming too humid, encouraging rot. Keep vents closed until seeds have germinated.

▲ WINDOWSILL PROPAGATOR
Portable propagators can be used indoors to maintain the moisture needed to germinate seeds. Some are fitted with heating elements, to provide bottom heat, and modular inserts.

▶ TENTING
The easiest way to cover a pot is to cover it with a clean, clear plastic bag held up with a wire hoop or a few split canes.

CHOOSING A GROWING MEDIUM

AN APPROPRIATE GROWING MEDIUM is crucial to success when raising plants under cover. Whatever is used, it must be free draining yet moisture retentive, well aerated, and free from pests and diseases. It may be a traditional medium, such as a good proprietary seed compost. Or it may be very modern, such as modules formed of rockwool, a material favoured by commercial growers. Some plants can even be raised on nothing more than moist paper.

COMPOST FOR GROWING FROM SEED

There is a wide range of composts available for use in propagation. Seed compost is more moisture-retentive and finer-textured than standard potting compost, to allow good contact between fine seeds and the moist compost, aiding germination. Seed compost is also low in nutrients, because mineral salts can harm seedlings.

Ready-made seed compost generally contains sterilized loam, peat substitute or peat, and sand. Soil-based composts are better for seedlings that will stay in their pots until well grown. Soilless seed composts contain no loam; these are suitable for seeds that germinate and grow quickly and are transplanted after a short time.

SEED COMPOST
This soilless compost contains perlite for drainage and a small amount of slow-release fetilizer.

USING COMPOST

- Buy and use compost as soon as possible; old, stale compost may not remain sterile. Store any unused compost in sealed plastic bags to prevent contamination.
- Do not sieve seed compost finely: if the texture is too fine it may form a crust (capping) through which the seedlings have to break. Sieve compost through your fingers or through a coarse sieve.

THE IMPORTANCE OF FIRMING COMPOST

Any growing medium must contain some air: lack of air leads to waterlogging and poor root growth. Large pockets of air in the compost, however, prevent water from being drawn up evenly to the roots.

Firm compost carefully; different composts should be firmed to different degrees. Lightly firm soilless compost, especially at the edges of a container; loam-based composts can be firmed slightly more.

FIRMING COMPOST
Water is drawn up through compost to the seedling roots by capillary action. Water cannot pass through air pockets, and growth will be uneven or poor. If the compost is properly firmed, air is distributed evenly and water is drawn up uniformly.

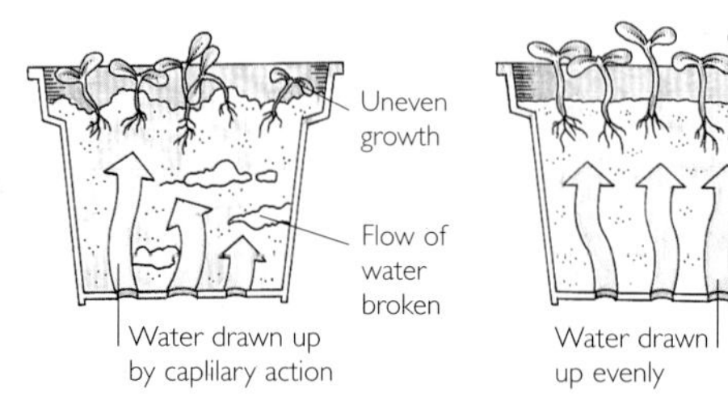

Alternative Growing Media

Even the best soil or compost can contain pests and diseases. There are a number of sterile, inert media available, all of which avoid these problems and also discourage damping off (*see p.43*). Commercial growers most commonly use rockwool, and this is an easy and widely available option. Rockwool is made from fibres spun from molten rock: do not confuse it with the water-repellent rockwool that is used in buildings. It allows air to the seedling roots and retains moisture. Seedlings can be potted on and planted out still in the module: rockwool disintegrates over time.

USING MODULES

- Squeeze a corner of the modules daily: if no moisture comes out, water them.
- If growing on in rockwool planting blocks, feed the seedlings with liquid fertilizer.
- When planting out, cover the rockwool well.

ROCKWOOL "CUBES"

Small rockwool cubes or modules have a hole for seeds

Large planting blocks have holes into which the smaller cubes are placed

ROCKWOOL MODULES
Soak modules first, then drain. Insert one or two seeds in each module. When roots appear, put the module into compost or larger planting blocks.

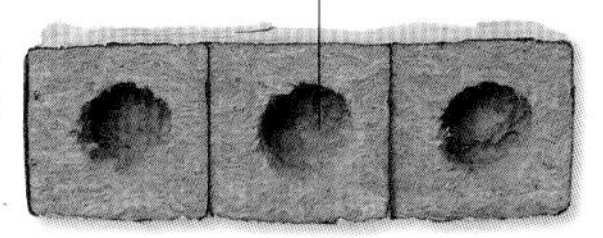

ROCKWOOL PLANTING BLOCKS

USING WATER AS A GROWING MEDIUM

Some salad crops can be grown using just water. Salad rape, mustard, and cress are easily grown simply on a wet kitchen paper towel and harvested when still seedlings. Chickpeas and some beans, such as mung beans, can be sprouted in a jar with muslin over the top, watered and drained daily, and similarly eaten when still young.

1 **Line a saucer** with kitchen paper and wet it. Scatter seeds over the paper. Cover with a clear plastic bag.

2 **Label and leave** on a warm windowsill, watering if needed. The seedlings will be ready to harvest in 7–10 days.

Compost Coverings

Coverings can help to prevent a cap or crust from forming on the surface of the compost of slow-germinating seeds. They hold small seeds in contact with the compost while discouraging rotting because they are open and allow air to get to the seed. They can also stop small seeds from being disturbed or dislodged when watered from above, because the water trickles gently through the covering.

TOP DRESSINGS
Several substances are suitable for covering the surface of pots or trays. Vermiculite is best if germination is likely to be rapid; grit can be used if seedling development is slower.

VERMICULITE

GRIT

COARSE SAND

Sowing Large Seeds

Large seeds are particularly easy to sow. What is meant here by large seeds is those that can be easily pinched between the finger and thumb and sown individually rather than shaken out or scattered. It is easiest to form a picture in the mind from a few well-known examples: among the woody plants, oaks, acers, and pines have large seeds, while among the annuals and perennials good examples include sweet peas, lupins, and marigolds.

Sowing in Individual Pots

Particularly large seeds, especially some tree seeds, or those that produce seedlings with long tap roots, can be sown individually in deep 10cm (4in) pots. This allows them to reach a good size and be potted on or planted out with the very minimum of disturbance to their root system.

SLOW-START SEEDS

- Many larger seeds germinate slowly. Some tree or shrub seeds may take up to two years.
- Cover pots containing slow-germinating seeds with a layer of fine grit or gravel.
- Remember to keep checking and watering.

SOWING ACORNS (OAK)
Press each seed into unfirmed, soil-based seed compost, and cover to its own depth with more compost to 5mm (¼in) below the pot rim.

Sowing in Containers

Where only small numbers of plants are to be raised, sowing into pots is ideal, taking up little space. Seedlings can be left to grow in pots for longer than they could be left in seed trays, because the greater depth allows longer roots to develop: plants would be held back if kept in the shallow confines of a seed tray. Large seeds of trees, shrubs, climbers, and perennials are most often sown in pots rather than trays.

1 **Fill an** 8cm (3in) pot with standard seed compost and firm gently to about 1cm (½in) below the rim. Sow seeds singly, spacing them evenly over the surface.

2 **Sieve seed compost** over the seeds until they are covered to their own depth with compost. Do not use a very fine-meshed sieve for this, or a crust may form.

3 **Cover** the compost with a 5mm (¼in) layer of small gravel, especially if they are tree or shrub seeds. Label and water well, then place in the appropriate environment.

Sowing in Modules

Seeds that germinate quickly and easily, such as delphiniums or lupins, or those of plants that dislike root disturbance, are best sown singly in module, or plug, trays (*see p.34*); use one with cells large enough for seedlings to reach a good size before they need to be moved on.

SOWING SINGLY
Fill a module tray with seed compost and firm very gently. Sow seeds singly and cover to their own depth with compost, vermiculite, or grit. Label and water.

Avoiding Disturbing the Root System

The growth of some plants, such as sweet peas or deep-rooted trees or shrubs, will be checked if the young roots are disturbed by transplanting. These plants need special pots. Root trainers are like a row of very deep modules, hinged at the side and opening to allow the root system to be lifted out. Tube pots and degradable pots also work.

◀ TUBE POTS
Simple tubes are excellent for sweet peas. Sow seeds singly, then cover with 1cm (½in of compost. Slide out the root system when transplanting.

▶ BIODEGRADABLE POTS
When the seedling is established, plant the whole pot; the roots will grow through the pot and it will rot away.

Sow up to three seeds and thin seedlings to one

USING SOILLESS ALTERNATIVES

Large seeds, especially those that have good germination rates, may be sown singly in rockwool modules (*see p.37*). Using rockwool reduces root disturbance, as the whole block is potted or planted, but for plants that have very long roots use the pots described above.

1 **Soak the rockwool** with tepid water and leave to stand for 30 minutes. Drain thoroughly; the modules must be moist, not waterlogged.

2 **Make a hole** about 5mm (¼in) deep in each module if necessary. Drop a single seed into each hole and cover it with loose rockwool fibres.

3 **Leave in a** warm, bright place. Check daily to see if they need watering. Put the modules into larger blocks or pots or plant them out.

Sowing Smaller Seeds

Smaller seeds are those that cannot easily be pinched between finger and thumb and planted singly. These are generally the seeds of many herbaceous perennials, annuals, and biennials. With all smaller seeds a prime consideration is not to sow too thickly, as seedlings that grow too close together tend to be spindly and prone to attack by fungal diseases. Some plants have seeds that are so fine as to be like dust, and these require particular care.

Sowing in Pots

Several kinds of seeds are best sown in pots, for a variety of reasons. Perennial seeds are often sown in pots or half pots of 9cm (3½in) to 13cm (5in) diameter, because some may not show signs of growth for up to a year after sowing. Cacti are often slow to germinate and so it makes sense to sow them in smaller containers. Alpines are generally sown in quite deep pots with plenty of drainage material in the base, as this provides the free-draining conditions that they require.

1 **Fill a container,** here a 13cm (5in) pot, with moist seed compost; loam-based is best for slow-germinating seeds. Firm it gently to 1cm (½in) below the rim.

2 **Sow the seeds** thinly and evenly, across the surface of the compost. This can be done straight from the packet, but you may find it easier from a folded piece of paper.

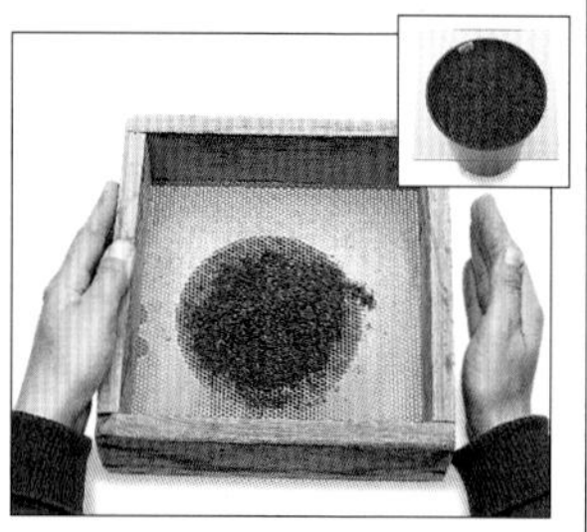

3 **Cover with a** shallow layer of compost. Stand the pot in water until the surface darkens; allow to drain. Cover with glass or plastic and stand in a warm, bright place.

HOW TO SOW DUST-LIKE SEEDS

Very fine, dustlike seeds require special treatment. These seeds should not be covered with compost, as they do not have enough reserves of energy to push up through a covering layer into the light. They are also liable to clump together and so grow too close, and are easily dislodged when watering. Water by standing the pot in water until the surface darkens and then allowing it to drain.

1 **Mix seeds** with fine, dry sand to make them easier to sow. Put the sand and the seeds into a small bag, seal, and shake well to mix.

2 **Fold a piece** of paper in half. Place some sand and seed mix on the crease and tap gently to scatter the mix evenly over the compost.

SOWING IN TRAYS

Plants raised in large quantities (generally annuals and biennials) are often grown in trays, as these make good use of space for large numbers. Seed trays are shallow and do not allow for much root development, so they are most useful for plants that are transplanted quickly into modules, pots, or the garden. Because trays have a large surface area, from which moisture can evaporate, you will need to keep a close eye on them to ensure that they do not dry out.

TRAYS AGAINST POTS

- Pots are better for small numbers of seeds, where using a tray would waste space.
- Trays are best where large numbers of seeds that germinate quickly are sown.
- Pots are better for plants that should grow to a good size before transplanting.
- Trays are not good for plants that dislike root disturbance, as seedlings raised in trays will always need pricking out (*see pp.44–45*).

1 Fill the tray with seed compost, mounding it up loosely above the rim. Tap the sides of the tray briskly to remove air pockets, then draw a piece of wood across the top to level. Firm gently to 1cm (½in) below the rim.

2 Scatter the seeds thinly and evenly over the compost surface. Just cover with sieved compost. Stand the tray in water until the surface darkens, then drain, or water with a fine rose. Cover with glass or plastic.

SOWING INTO MODULES

Smaller seeds can be sown into modules, several to a cell. The resultant seedlings can either be thinned, or, as is sometimes done with vegetable crops, planted out as a block. This is known as "multiblock sowing", and is used particularly for root, bulb, and stem vegetables.

CROPS IN MODULES

Beetroot ***(Beta)*** Sow one seed per module, as most cultivars have multigerm seeds.
Carrots ***(Daucus)*** Round-rooted cultivars of carrot may be multiblock sown.
Onions ***(Allium)*** Sow six seeds per module and grow as a group.
Turnips ***(Brassica)*** Multiblock sowing is used to produce golf-ball-sized "baby" turnips.

MULTIBLOCK SOWING
Fill a module tray with moist compost and make a shallow depression in each module. Sow a few seeds in each and lightly cover with compost. Plant out the seedlings as a block.

CARE OF SEEDS AND SEEDLINGS

GENERALLY, SEEDS REQUIRE WATER, warmth, air, and sometimes light in order to germinate successfully. Once the seedlings appear, they need water, warmth, air, light, and nutrients in order to continue growing. Meeting these requirements in the correct amounts under cover takes a little care, but good observation and attention to a few basic principles are all that is necessary to raise a very broad range of plants.

MOISTURE AND DRAINAGE

Seeds need water to grow, but it is vital that they should not be waterlogged or kept in an environment that is too humid; this can lead to "damping off" (*see opposite*). A tent of plastic, sheet of glass, or propagator lid over a pot or tray of seeds will keep in moisture; remove it as soon as the seedlings appear and monitor water levels carefully.

WATERING
If you use a watering can to water seeds and seedlings, always choose a very fine rose, so that the spray is gentle.

CLEAR COVERING
A plastic bag over a pot of seeds helps to conserve moisture and removes the need for frequent watering.

TOP DRESSING
A dressing of grit will ensure that water drains away freely from the bases of seedling stems.

LIGHT LEVELS

It is vital that seedlings receive the right amount of light. If a seedling does not get enough light, it will become pale, with a weak stem and small leaves that are widely spaced. Always keep seedlings in a bright, airy place. Spindly seedlings can be planted deeply when they are pricked out (*see pp.44–45*), with most of the long stem under the surface.

Strong direct sun can scorch leaves, so shading may be necessary for plants sown at warmer times of year. Shading material should allow enough light for good growth to pass through it. Shading washes are used for the greenhouse; shading netting can be supported over pots or trays. In warmer climates, structures made from timber slats are useful for creating dappled shade.

◄ SHADING A GREENHOUSE
Washes on a greenhouse allow light through, but reduce the scorching heat of the sun, or blinds can be used to shade plants as required.

► SHADING A SINGLE TRAY
Netting shades seeds from direct sun until the seedlings appear, and is then removed.

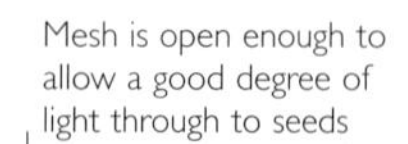

Mesh is open enough to allow a good degree of light through to seeds

Preventing Damping Off

The most common problem with seedlings raised indoors is "damping off", caused by fungi that spread rapidly in wet compost and humid, warm conditions. Seedlings flop over, often with a brown, shrunken ring at the stem base, and white fungus appears. Keep everything clean when sowing seeds (*see p.35*), and avoid poor light and dense sowings, which increase the problem. Treat with a copper-ammonium fungicide.

◀ DAMPING OFF EVIDENCE
The warmth and moisture needed for germination and seedling growth also encourage the growth of fungi.

▶ PREVENT BY SPRAYING
Fungicides applied on sowing and when the seedlings emerge can prevent problems.

Greenhouse Pests

Although pests can appear in plants on your windowsills, they are more often found in the greenhouse. To avoid an infestation of red spider mites, woodlice, or sciarid and whiteflies during the growing season, scrub the propagation area annually with a solution of horticultural disinfectant. This also helps to control mildew, and the fungi that cause damping off (*see above*). Slugs and snails can be removed by hand, poisoned with pellets, drowned in beer traps, or kept out by barriers.

SNAIL

SLUG

COMMON PESTS
Young seedlings are soft and particularly vulnerable to attack by garden pests. Keep the propagation area clean and tidy to discourage infestations.

WOODLICE

INSECT TRAPS
Yellow sticky traps hung above plants will attract and trap many insect pests, including whitefly.

PROBLEMS AFFECTING SEEDLINGS

Sciarid fly or fungus gnat These greyish-brown flies, 3–4mm (⅛in) long, fly or run over compost, and the larvae feed on the roots of seedlings. Predators or pesticides can be used to kill them, but try to avoid infestations by keeping materials scrupulously clean when sowing and not overwatering seedlings.

Vine weevil The plump, creamy white grubs of the vine weevil live in soil or compost and feed on roots. The outer tissues of seedlings, particularly those of woody plants, may be gnawed from the stems below ground, and plants wilt and may die. Grow seeds in fresh compost, away from other container-grown plants to help avoid infestation: pathogenic nematodes or compost containing imidacloprid kill the grubs.

Outgrowing the container If seedlings are not pricked out at the optimum time (*see pp.44–45*) they are prone to various problems, including overcrowding, which results in weak growth. They may also run out of nutrients, indicated by yellowing or reddish foliage and stunted growth.

PRICKING OUT AND GROWING ON

SEEDLINGS, WHEN GROWING STEADILY, will need transplanting or pricking out into further containers to give them room to grow. Seedlings from seeds sown singly in individual containers or modules, including root trainers, do not need pricking out: these seedlings may simply need potting on or planting out. Generally, pricking out takes place as soon as seedlings can be handled easily by the seed leaves, because overcrowded seedlings soon deteriorate.

PRICKING OUT FROM POTS

Once germination occurs, most seedlings should be transplanted as soon as they are large enough to handle by the seed leaves. Plants sown in deep pots may be left a little longer as they will have a little more space and nutrients to develop roots than seedlings in trays. They should not be left very long, however, as their developing root systems will become harder to tease apart, risking damage. Bulbous plants do not have this problem, of course, and are often too delicate to be transplanted in their first season: leave until the next year.

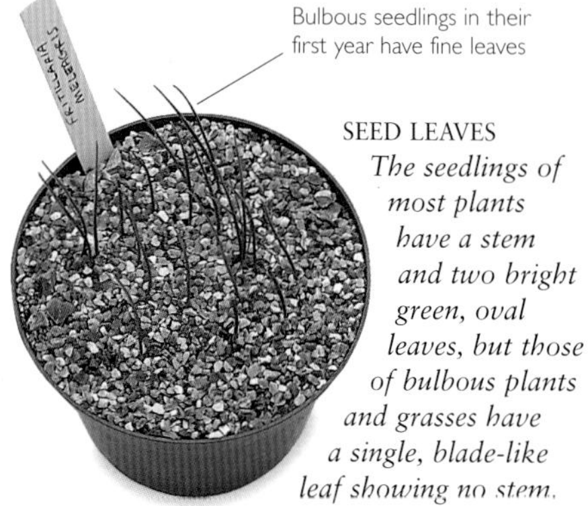

SEED LEAVES
The seedlings of most plants have a stem and two bright green, oval leaves, but those of bulbous plants and grasses have a single, blade-like leaf showing no stem.

1 Seedlings of most plants, such as summer bedding or perennials, are pricked out as soon as they can be handled easily, but those of trees and shrubs, here of birch (*Betula*), can be left until they are about 2.5–5cm (1–2in) high.

2 Knock the young seedlings out of the pot. The compost should break up, making it easier to tease out the roots. Always hold the seedlings by their leaves, as their roots and stems are still very fragile and are easily damaged.

3 Transplant each seedling into its own 8cm (3in) pot filled with standard potting compost. Firm gently around the seedling, label, and water. Grow on in the same place as before. Harden off seedlings gradually after 3–4 weeks in the new pots.

PRICKING OUT BEDDING PLANTS

Container-raised seedlings should be transplanted into larger containers as soon as possible, so that they have room to develop before being planted into their flowering positions. Seedlings of plants destined for bedding out are often pricked out into a deeper tray. The seedlings grow on in these trays until it is time to plant them out, so they must be given adequate space for subsequent development.

1 **Water the container** and allow it to drain, then tap it on a hard surface to loosen the compost. Ease the seedlings out, or lift the whole mass out of the tray and tease apart.

2 **Place the seedlings** about 2.5cm (1in) apart in a prepared tray of potting compost, handling only by the leaves and using a dibber or similar tool to make holes. Water and label.

PRICKING OUT INDIVIDUALLY

Plants to be grown on to a moderately large size before potting or planting out will suffer less of a check to their growth if they are transplanted as soon as they can be handled, even if they are quite small. Seedlings grown on in modular trays are easy to handle and suffer little check to growth when transplanted to their final position. Other containers for growing on individual plants are biodegradable pots for minimum root disturbance, or plastic pots for plants that will be kept in containers.

1 **Water and drain** the tray, then gently knock the compost out. Lift each seedling away, handling only by the leaves and using a widger or a small spoon under the roots.

2 **Transplant each** seedling into a prepared container (here a module), making a hole large enough for the roots. Gently firm the compost around the seedling. Water and label.

HARDENING OFF AND PLANTING OUT

NEW PLANTS RAISED UNDER cover in cool climates will have fairly soft growth, and must be acclimatized to temperatures outdoors, or "hardened off', over at least three weeks. During hardening off, the natural waxes coating the plants' leaves undergo changes in their form and thickness to reduce water loss. The pores on the leaf, through which water evaporates from the plant, also need to adapt to the less favourable conditions outdoors.

WHEN TO BEGIN HARDENING OFF

Hardening off should start when the plants are established in their containers, several weeks before they are planted out in the garden. Most plants need at least three weeks of gradual adjustment to outdoor conditions, and even up to six weeks is not too much.

In cool and cold climates tender plants, such as summer bedding plants, must not be planted out until risk of frost is over. The hardening-off process for hardy plants raised under cover can start as soon as they are established and growing steadily.

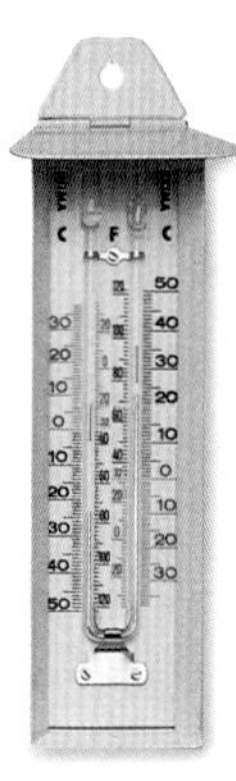

◀ DAILY TEMPERATURE
Thermometers can be bought that record the minimum and maximum temperatures with markers that can be reset daily. Use these to determine whether nights are still frosty.

▶ READY TO GO
These young plants are ready to be hardened off. Their small rootballs make them vulnerable to frost, and the process must be gradual.

USING A COLD FRAME

Cold frames – fixed wooden frames with hinged glass or plastic lids – provide a halfway house between the greenhouse and the open garden in cool climates. The soil and air temperatures in a cold frame are higher than in the garden and fluctuate less. Cold frames also provide shelter from winds while ensuring adequate light levels.

Transferring young plants to a cold frame for hardening off is ideal. The frame can be ventilated increasingly, as conditions permit, until the covers are fully open at night as well as by day. A cloche may also be used, but does not give as much frost protection as a cold frame.

HALFWAY HOUSE
A large, deep cold frame allows many plants to be hardened off. A thick layer of gravel in the base provides good drainage.

HARDENING OFF OUTDOORS

The simplest way to harden plants off is to place them in the shelter of a wall or hedge and cover them at night, and by day in poor conditions, with newspaper, plastic sheeting, or shade netting. Keep them out of direct sunlight, which could scorch them, especially in the middle of the day.

POINTS TO CONSIDER

- Keep an eye on the weather and remember to cover young plants at night until the end of hardening off.
- Water young plants well in warm weather; they may be losing more moisture outdoors.
- Once plants are outside they will need protection from outdoor pests (*see p.59*).

TEMPORARY COVER
A simple tunnel of plastic sheeting over hoops is enough to protect many seedlings. Raise the sheeting to allow more ventilation each day, rolling it back down at night.

PLANTING OUT

Hardened plants may be transferred into their final positions in the garden once the conditions outdoors are right. For frost-tender and half-hardy plants, this is not until the last expected frost has passed; for hardy plants the timing is less critical.

When planting out, it is important to take into account the eventual size of the plants and space them accordingly; small plants from containers grow quickly once in the open ground. Also important is the correct depth for the plant. Most plants should be planted at the same depth as they were in the pot, but plants that prefer moist conditions may need to be planted a little deeper, and plants that are liable to rot if too moist may need to be planted out with their crowns slightly above the soil level.

PLANTING OUT FROM MODULES
First water the trays well. Each plug should come out with a good, clean rootball; modules have a hole in the base, so use a piece of wood or cane to push the plugs out. Firm in, just covering each rootball, and water well.

PLANTING OUT LARGER PLANTS
Water the pot thoroughly and allow to drain. Gently slide the plant out, taking care not to damage either the roots or the growing points. Put the plant in the hole, fill around it with soil, firm with your fingertips, and water in.

Growing Ferns from Spores

Ferns lack flowers, reproducing by spores rather than by seeds. Increase from spores is the usual method of propagation for producing lots of plants. The fern life cycle has two phases: the familiar, fronded, plants we grow shed spores, and when these land on the soil they develop into a green pad called a prothallus. If the prothallus is kept moist, male parts on it fertilize female parts; an embryo develops, then a recognizable fern, which will in turn produce spores.

Selecting a Ripe Frond

Spores of most temperate fern species ripen in mid- to late summer; those of many tropical ferns ripen less seasonally through the year. The spore-bearing bodies are usually visible as small patches on the underside of the fronds; a few ferns produce special spore-bearing fronds in the centre of the plant. As these bodies ripen, their colour darkens and they shed the dust-like spores.

UNRIPE FROND
Unripe bodies are usually pale green or pale brown in colour, and have a granular surface. Leave these to ripen.

RIPE FROND
When just a few of the bodies are open and are shaggy in appearance, the frond is ready for propagation.

OVER-RIPE FROND
When most of the bodies have turned brown and shaggy they are over-ripe, and most of the spores have been shed.

Collecting Spores

To collect spores, place a spore-bearing frond or section of frond in a clean envelope and keep in a warm, dry atmosphere. Do not use plastic bags; they encourage moulds. When the spores are released, they have the appearance of fine dust. Before sowing, they should be separated from any debris such as scale remnants or leaf hairs, which can contaminate the spore culture.

REMOVING A FROND
Cut off the frond with a clean, sharp knife. Place it in a clean folded sheet of paper or envelope in a warm, dry place for 2–3 days to collect the spores.

SOWING THE SPORES

The minute spores can be separated from the debris either with a fine sieve, or by tipping them onto a clean sheet of paper and holding it at an angle of 45°. Debris will slide down the surface faster than the spores; with a little practice, the spores can be kept on the paper while the debris falls off.

Green spores must be sown within 48 hours of collection. Spores that are brown when ripe can be stored in a refrigerator at 4–5°C (39–41°F) with a sachet of silica gel.

EASY FERNS FROM SPORES

***Athyrium* (Lady fern)** Sow fresh spores at 15°C (59°F).
***Blechnum* (Hard fern)** Sow spores in late summer at 15°C (59°F).
***Dryopteris* (Buckler fern)** Fresh spores germinate readily at 15°C (59°F).
***Matteuccia struthiopteris* (Ostrich fern)** Germinate fresh spores at 15°C (59°F).
***Osmunda regalis* (Royal fern)** Sow green spores as soon as ripe at 15°C (59°F).

1 **Sterilize a mixture** of equal parts peat and sharp sand, or two parts sphagnum moss peat to one of coarse sand, by pouring boiling water over it. Put in an 8cm (3in) pot. Gently tap the spores onto the surface from a clean piece of paper. Cover with clear kitchen film.

2 **Keep the pot** in a closed propagator in indirect light and at the appropriate temperature for the species. After 6–9 months you can carefully lift small "patches" of the green prothalli that have developed on the compost surface.

3 **Set the patches** up to 2cm (¾in) apart in a pot of fresh compost, placing them in slight depressions in the compost surface. Spray the patches with cooled boiled water, cover, and place the pot in the same propagating environment as before.

4 **When the young** fronds are large enough to handle, pot them into modules or trays of moist, soilless potting compost. Keep these in a humid environment and pot the ferns on when larger fronds develop. Most new ferns are large enough to plant out in 2–3 years.

SOWING DIRECT

MANY SEEDS ARE BEST sown outside. Seeds that germinate slowly can rot if their compost becomes sour, so these may be better sown into a nursery bed in a cold frame. Hardy biennials and perennials can be grown in a seedbed; hardy annuals and most vegetables are generally sown straight into their final position. Success depends on good preparation of the ground: it should be clear of weeds, have a good open structure, and contain plenty of organic matter.

CLEARING THE GROUND

It is vital that the soil into which you sow is free of weeds that would smother seedlings. If you are creating a new sowing area, the first step is to clear away all weeds. Annual weeds can be removed by hoeing, perhaps together with the stale seedbed technique (*see below*); perennial weeds are tougher and require either meticulous hand weeding or treatment with a weedkiller.

PERENNIAL WEEDS

- Larger areas can be cleared with glyphosate weedkiller, which breaks down in the soil.
- If you prefer not to use weedkiller, repeated removal of weeds or a light-excluding mulch can be used; these are slower methods.
- If weeding by hand, remember that for some weeds even a fragment of root left in the soil can make a new plant, so remove everything.

FORKING OUT WEEDS
If sowing direct into a border you will first have to clear out any weeds and the remains of previous plants. Use a hand fork or a border fork to tease out weed roots without creating too much disturbance among any established garden plants.

THE STALE SEEDBED TECHNIQUE

This technique helps to destroy as many weeds as possible before sowing seeds in a seedbed. Any garden soil contains dormant weed seeds. Cultivating the ground a few weeks before sowing brings these weed seeds close to the surface, giving them the air and light that they need to germinate. The weeds can then be hoed or sprayed with weedkiller; it is important not to further disturb the ground at this stage, as that would bring up yet more weed seeds.

1 **Dig over** the sowing area lightly to disturb any dormant weed seeds that are in the top layer of the soil. Leave for a few weeks.

2 **Weed seeds** will germinate on the cultivated ground. Clear them by light hoeing or with a weedkiller, taking care not to disurb the soil.

Digging and Forking the Soil

Dig soil over in autumn, adding organic matter, such as well-rotted manure or garden compost, if necessary. When you come to sow in spring, this will have been broken down by soil organisms and the nutrients will be available to the young plants.

Heavy clay soils may be easier to work if roughly dug in autumn and left for winter frosts to break up; on sandy soils it may be best to leave organic matter on the surface over winter and then fork it in. In spring, loosen the soil and remove any weeds.

SINGLE DIGGING
Dig a trench, 30cm (12in) wide and a spade's blade deep. Dig a second trench, placing the soil from it into the first trench. Continue, filling the last trench with the soil from the first.

FORKING THE SOIL
At a time when the soil is moist but not waterlogged, work over the area to be sown methodically with a fork, inserting it and turning it over to break up compacted soil into a crumbly structure.

IMPROVING THE SOIL
You can improve the structure and fertility of the soil by incorporating organic matter such as well-rotted manure or garden compost. Fork this in during or after digging or when forking the ground.

Growing in Beds

Vegetables have traditionally been grown in spaced rows, or "drills", in rectangular plots; nowadays the bed system, with crops spaced equally in narrow beds lined by paths, is more popular. The benefit of this system is that only the actual bed is dug, manured, and fertilized, not the soil in-between. Also, all of the work can be done from the paths, avoiding soil compaction.

SOWING IN RAISED BEDS

- Plants are spaced equally in all directions, because there is no need to leave space to walk between rows.
- Raised beds warm up more quickly in spring, allowing earlier sowing.
- It is possible to sow over four times as densely in a bed as in a conventional plot.

MAKING A RAISED BED
Mark out the area of the bed: it should be no more than 1.5m (5ft) wide to allow access without treading on the soil. Cultivate the bed, digging in organic matter and mounding the surface with topsoil from the path area.

Preparing the Surface

Just before sowing, remove weeds and firm and rake over the soil to give a smooth, loose surface, known as a "fine tilth". Heavy, wet soils are cold and lack oxygen: if possible, wait until the soil is workable before sowing or transplanting seedlings. If the soil is wet, stand on a board to avoid compaction. The soil should not be dry, since water is needed to soften the seed coats and start germination.

FIRMING AND RAKING

- Firming the soil removes any large air pockets that would make the flow of water through the soil uneven.
- Raking to produce a fine tilth allows seeds to be sown at a consistent depth and provides fine soil for covering them.
- Raking helps when sowing seeds broadcast in particular, as the seeds settle into the small furrows on the surface.

1 **Firm the ground** by shuffling forwards with both feet together over the whole area until it is flat. Pay particular attention to edges.

2 **Rake over** the area in all directions to create a fine tilth, ready for sowing. If the soil is dry, water it thoroughly.

Sowing Broadcast

This method is best used when sowing among other plants, for example when filling gaps in borders, and it is most often used for annuals and biennials. Some vegetables, such as carrots and radishes, can also be sown broadcast; this makes efficient use of space and may be used for early sowings into a cold frame or in a plastic-film tunnel in cool climates. Weeding broadcast sowings can be more difficult in the early stages, as a hoe cannot be used, and distinguishing between sown seedlings and weed seedlings may be tricky at first. For this reason it is best to use the stale seedbed technique (*see p.50*) before making broadcast sowings. Sow the seeds thinly on the prepared surface and rake them in lightly to keep them in contact with the soil. Always water in with a watering can with a very fine rose to avoid disturbance.

1 **Use a rake** to give the soil a fine tilth and leave fine furrows on the surface. Scatter the seeds thinly and evenly over the prepared seedbed either by hand, from the packet, or with a seed sower.

2 **Cover the seeds** by raking over the area at right angles. Use light strokes, so that the seeds are disturbed as little as possible. Label the area and water the soil using a fine-rosed watering can.

Sowing in Drifts

This technique is generally used for sowing annuals in bold, informal groups. Make a plan before sowing, giving consideration to height, spread, habit, and flower colour. Divide the area into a grid to help transfer the plan to the ground accurately.

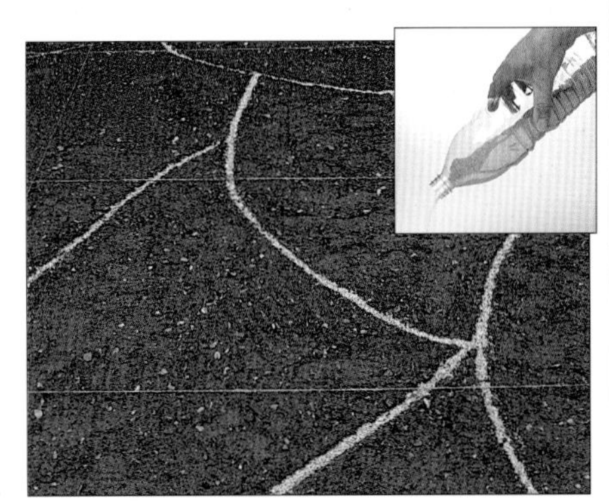

MARKING OUT AREAS
Use canes or twine to mark out a grid on the sowing area, then transfer the plan to the ground. Mark the sowing areas with a stick or by pouring fine sand from a bottle. Sow each area broadcast or using mini-drills (see below).

Sowing in Mini-drills

Because broadcast sowing can result in patchiness and make weeding difficult, it is often better to sow in "mini-drills". These initially look more formal, but the rows do blend eventually, and they are easier to weed and thin. Drills should be 8–15cm (3–6in) apart, depending on the eventual size of the plants, and no more than 2.5cm (1in) deep. The depth should be uniform along the drill to ensure even germination.

1 **Mark drills** by scoring the surface with a stick or the corner of a draw hoe, or by pressing a cane or the back of a rake into the soil. Make the drills in adjoining areas run in different directions.

2 **Scatter the seeds** thinly and evenly along the length of the drill. Space large or pelleted seeds individually. If the soil is very dry, water each drill before sowing. Label each area as it is sown.

3 **Carefully rake** the soil back over the drills without dislodging the seeds. Firm with the back of the rake and water the area using a watering can fitted with a fine rose.

4 **As the seedlings** grow the pattern of rows makes it easy to distinguish plant seedlings from weed seedlings, and it may even be possible to hoe between widely space drills. The regimented rows will quickly become obscured as the young plants grow.

SOWING RATES

- Sow old seeds more thickly, as the germination rates will be lower.
- It is better to sow more seeds and thin the seedlings than to sow too sparingly and get patchy results.

Sowing in Standard Drills

Standard drills are generally used for sowing vegetables. Most vegetables prefer a free-draining, moisture-retentive, slightly acid soil and one that is rich in nutrients. When digging over the soil (*see p.51*) add plenty of organic matter, but do not sow any root crops (apart from potatoes) on freshly manured ground, because they will produce forked roots. In spring, loosen up the soil and add fertilizer. Firm, rake over the soil to give a fine tilth (*see p.52*), then mark out drills at the correct spacing.

1 Mark out a row with a string line and pegs, or with a cane. Use the corner of a hoe to draw out small, even drills in the soil to the depth and spacing that are required for the seeds.

2 Stand on a plank to avoid compacting the soil when sowing. Sprinkle the seeds thinly and evenly along the drill. Cover the seeds with soil, without dislodging them, and water in.

Using Drills

- When the soil is very dry, water the base of the drill first, then sow the seeds and cover over with dry soil.
- If the soil drains slowly or is very heavy, sprinkle a layer of sand in the bottom of the drill before sowing the seeds.
- Two crops may be sown in the same drill to maximize use of the ground. A fast-growing crop is sown in between a slow-growing crop and harvested before the slower crop fills the space.

Sowing in Wide Drills

Wide drills with flat bottoms are used for crops that are grown close together, such as peas, early carrots, and crops that are harvested when still seedlings. Drills should be 15–23cm (6–9in) wide, to allow the seeds to be spaced correctly for their size. The correct depth will vary according to the seeds, but the depth must be even along the drill.

1 Take a hoe and drag it towards you, applying a light and even pressure. Mark out parallel drills at the required depth for the seeds.

2 Space large seeds, or trickle-sow smaller seeds, along each drill. Make sure that the required distance is left between the seeds.

3 Carefully cover over the seeds with soil using the hoe or a rake and taking care not to dislodge the seeds. Water in well.

Space-sowing at Stations

This method reduces thinning, makes economical use of seeds, and avoids the need to transplant crops. Drills are made at the correct spacing and depth for the crop. The "stations" at which to sow are marked by drawing out more drills or by making shallow holes along each drill.

MARKING STATIONS
Draw out drills at correct spacings, then draw more drills at the correct spacing at right angles to the first set. Sow 2–3 seeds at each intersection or "station". Water in and label.

Sowing Large Seeds in Rows

Large seeds, such as those of beans and sweet corn, may be sown individually in holes made with a finger or a dibber. Make sure that each seed is at the base of the hole and in contact with the soil. Cut-off plastic bottles or glass jars can be used to protect tender sowings until the seedlings emerge.

ROWS OF BEANS
Beans have large seeds that germinate reliably, and so can be planted individually. Use a cane or line as a guide. Drop a seed into each hole, cover with soil, water, and label.

Fluid Sowing of Pregerminated Seeds

Seeds of some crops, such as beetroot and parsnips, can be pregerminated (*see p.33*) and then fluid-sown. Sow when the seed roots are no longer than 5mm (¼in). Mix the seeds with a clear gel, such as wallpaper paste: do not use a paste containing fungicide. Gel helps to keep seeds moist, but they may still need watering.

1 **Pregerminate** the seeds. As soon as they have begun to sprout, wash them carefully into a fine-meshed sieve under gently running water.

2 **Mix up** wallpaper paste (without fungicide): about 250ml (8fl oz) for 100 seeds. Tap the seeds into the jar and stir gently to distribute them.

3 **Pour the paste** into a plastic bag and knot the end. Snip off a corner and gently squeeze a line of paste into the drill. Cover the seeds and label.

Sowing in Pots Outside

Hardy plants raised in substantial numbers are often sown in outdoor seedbeds (*see pp.50–55*), but some subjects are better sown in pots. This applies especially to seeds that germinate slowly or erratically, and those that need cold stratification over winter. The hardy plants usually germinated in pots outdoors include trees, shrubs, climbers, perennials, alpines, and bulbs. Some vegetables may also be started this pots.

Sowing Seeds in Pots

Always use a good quality, gritty seed compost containing only a little fertilizer, as too much can kill seedlings, especially if they remain in their pots for some time after germination. The compost must be well drained, otherwise the seeds may rot as a result of sitting in wet conditions, but at the same time it must be able to retain sufficient moisture for the seeds to germinate. Seeds sown outdoors are liable to be disturbed and eaten by pests such as mice and birds, so when the pots have been placed in a cold frame or cool spot outdoors (*see right*) cover them with fine-mesh wire netting, tucking it under the pots to deter rodents.

1 **Prepare a pot** with seed compost and firm. Sow the seeds evenly across the surface of the compost from the packet, your hand, or a folded piece of paper.

2 **Use a soil** sieve to scatter a thin layer of compost over the seeds, covering them to the appropriate depth. Water gently with a fine rose or stand the pot in water.

3 **Cover the compost** of seeds that are slow to germinate with a layer of 5mm grit. Fill to just below the rim, adding the grit carefully to avoid disturbing the seeds.

4 **Label the pot** then either stand it in a shaded nursery area or plunge it in a sand bed (*see above right*), which will help to prevent the compost from drying out.

THE VALUE OF USING TOP DRESSINGS

If pots are standing outdoors for a long time, moss and liverworts that compete with seedlings may grow on the surface. It is best to cover seeds with a top dressing to inhibit these growths. It will also make it harder for weed seeds to settle, and discourage slugs from attacking young seedlings.

GRIT COVERING
Cover the compost in pots outside with 5mm grit or coarse sand. This is particularly important for seeds that germinate slowly, as the compost will otherwise be exposed for a long time.

Overwintering

Some seeds, especially of hardy shrubs or climbers, require a period of winter chilling or "cold stratification" before they will germinate; this can be done before sowing (*see p.31*) or after, by overwintering seeds sown in autumn outside or in a cold frame.

Plunge beds of coarse sand or grit help retain moisture and moderate compost temperature, so that it does not freeze in winter – or overheat in summer. Clay pots absorb water from the sand or grit; water plastic pots directly but sparingly.

STRATIFYING
In cooler climates, place pots in a cold frame after autumn sowing. The frame allows exposure to cold, while protecting the seeds.

PLUNGE BED
Seeds that are sown fresh can be plunged in an outdoor sand bed or cold frame over winter. Cover the compost with a layer of grit.

Outdoor Nursery Beds

Large numbers of new plants and seedlings in containers can be grown on in an outdoor nursery bed. The beds isolate young plants from soil-borne diseases, and enable containers to drain freely while giving plants access to water through capillary action. Plastic linings keep out both weeds and disruptive worm activity.

FABRIC BED
If the soil is uneven or badly drained, cover it with sand. Line the bed with permeable membrane.

Woven or perforated fabric supresses weeds

Run the lining up sides of boards

Two way flow of water through lining

Sand improves drainage from bed

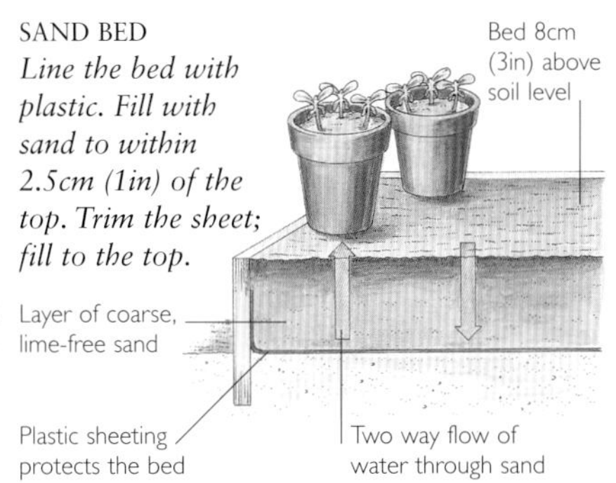

SAND BED
Line the bed with plastic. Fill with sand to within 2.5cm (1in) of the top. Trim the sheet; fill to the top.

Water-garden Plants

Seeds of aquatic plants should be sown immediately upon collection; if this is not possible, store them in phials of water.

Sow them in aquatic compost, loam-based potting compost, or sieved garden soil; do not add fertilizer, as it will encourage algae.

1 Collect ripe seeds in summer or autumn. Sow evenly over the surface of a 13cm (5in) pot of firmed compost. Cover with a 5mm (¼in) layer of fine grit to retain moisture.

2 Stand the pot in a large bowl a little deeper than the pot. Add water until it just covers the pot. Place in bright light, outdoors for hardy plants, until the seeds have germinated.

Care of Outdoor Seedlings

Outdoor sowings do not have the controlled environment found under cover, and seeds and seedlings may need protection. Winds can increase moisture loss: erect windbreaks on the side of the prevailing winds or use cloches, which also protect seedlings against the cold. In warm climates or seasons, beds may need watering and plants may even need protection from the sun. Some pests and diseases are inevitable, but damage can be limited with a little care.

Protection from the Cold

In the open garden in cooler climates, seedlings may need protection against cold. Cloches or tunnels are available in a wide range of designs and materials. They come in glass or various plastics: glass allows more light penetration and retains more heat, but is heavy and often expensive. Rigid plastic retains heat better than single-thickness plastic film, but film is cheaper. Plastic film or rigid polypropylene will last for five years or more, while rigid, twin-walled polycarbonate lasts for at least ten years. Horticultural fleece can also be used, or recycle bottles to make individual cloches.

◄ BOTTLE CLOCHE
Make an individual cloche by cutting the bottom off a clear plastic bottle. Leave the bottle top on and use it as a vent.

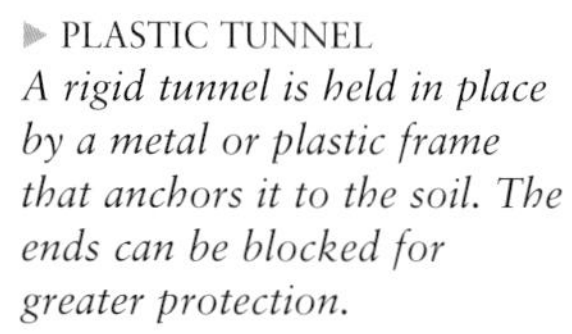

► PLASTIC TUNNEL
A rigid tunnel is held in place by a metal or plastic frame that anchors it to the soil. The ends can be blocked for greater protection.

PLASTIC BARN CLOCHE
The extra height of the sloping top makes this a versatile cloche. Many designs are available; large cloches will straddle a bed, as here.

FLOATING CLOCHE
Made of perforated plastic film or fleece held down by bricks, this simple, inexpensive system allows through light, air, and moisture.

FLEECE TUNNEL
A fleece tunnel is made from wire or plastic hoops and a sheet of horticultural fleece. The fleece is tied at each end and pegged into the ground.

Screening from the Sun

Seedlings outdoors will be exposed from the start to natural light. Early in the year in cooler climates this is not a problem, but later sowings or warmer climates bring a risk of the young leaves being scorched by the heat of the sun.

Protect vulnerable seedlings with shade netting held on hoops: this may only need to be in place for part of the day.

FLEXIBLE MESH
This can be cut to size and held over young plants on hoops, like a tunnel cloche.

Dealing with Weeds, Pests, and Diseases

Plants in the garden have to compete with weeds for space, light, water, and nutrients. Weeds can also harbour diseases, and it is vital to control or eradicate them around young plants. Regular weeding by hand is the best method; a hoe can be used between plants sown in drills (*see pp.54–55*) but it will not eradicate perennial weeds.

Pests range from rabbits and slugs to microscopic eelworms. Barriers will protect against some pests: cotton strung across seedbeds deters birds and barriers of mesh or fleece will stop rodents or carrot root fly. Plant diseases are sometimes carried by pests, so protecting against pests also provides some protection against disease.

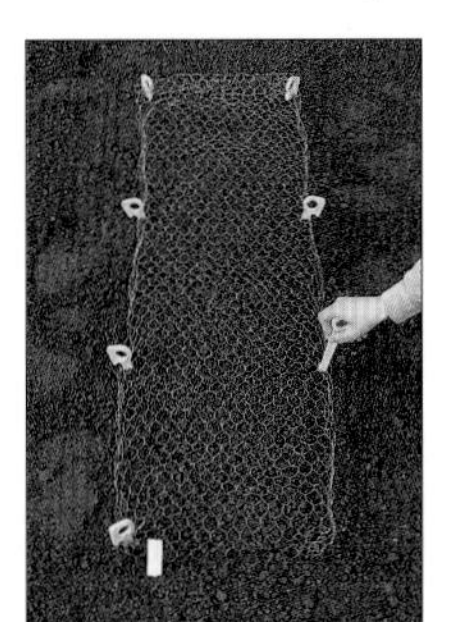

◀ BIRDS AND MICE
Protect seeds from birds or foraging animals if necessary by pegging wire netting over the row. Remove the netting before the seedlings grow through the mesh.

▶ FLEA BEETLE
Seedlings of vulnerable plants, such as brassicas, should be treated against flea beetle when they reach the two-leaf stage, using derris dust.

RECOGNIZING COMMON PROBLEMS

Ants Tunnelling may disturb seedlings. Use ant powder.

Aphids Small, green, sap-feeding bugs. Use an organic spray (derris or pyrethrum).

Cutworms Green-brown soil-dwelling caterpillars that eat roots. Search for and remove by hand.

Frost damage Leaves wilt, dry up, and turn brown. Nip off affected leaves.

Fungus gnats White maggots with black heads eat roots. Use a predatory mite or insecticide.

Grey mould (*Botrytis*) Grey fungal growth on leaves. Remove any affected parts of the plant.

Millipedes Have long, segmented bodies and can damage seedlings. Remove by hand.

Powdery mildew White, powdery fungal growth on leaves. Control with an organic fungicide.

Rusts Orange or brown pustules form under leaves. Remove infected leaves.

Scorch Leaves become dry, crisp, and brown. Protect from sun and wind.

Slugs and snails Eat foliage and stems. Use slug pellets.

Thrips Fine silver mottling on leaves. Use organic insecticide (pyrethrum or derris).

Viruses Distorted growth, leaves mottled or streaked yellow. Destroy the plants.

Wireworms Orange-brown soil-dwelling beetle larvae that chew seedling stems. Use insecticidal dust.

Thinning and Transplanting

Even with the most careful sowing, seedlings need thinning to allow the strongest to develop without competition from the weaker ones. It might seem less laborious or more economical simply to sow more thinly, but this would leave gaps, as some seeds will fail, and with smaller seeds it may not be possible at all. Many garden plants, particularly hardy annuals, shed copious amounts of seeds, so self-sown seedlings may also need thinning.

Thinning Seedlings

Whether you broadcast or sow in rows, seedlings will need thinning to avoid overcrowding. Use the strongest thinnings to fill in gaps caused by uneven sowing or poor germination, or grow on elsewhere in the garden. Thinnings of vegetable crops, such as carrots, can often be used in the kitchen as "baby" vegetables.

When to Thin

- Do not thin until seedlings are established and growing away well; this avoids leaving gaps where some seedlings fail after thinning.
- The best time to thin is when the soil is moist and the weather is mild.
- If the final spacing is to be 20cm (8in) or more, thin in stages to avoid any gaps being created if some seedlings die in the meantime.

THINNING CLUMPS
Seeds sown broadcast may come up unevenly, with some dense clumps. Lift clumps of seedlings and separate them, retaining plenty of soil around the roots. Replant the lifted seedlings singly into the bed at even spacings.

THINNING ROWS
To thin seedlings growing in drills, press down on the soil around the strongest seedlings with your fingers while pulling out the unwanted, weaker ones from around them. Refirm the soil and water.

THINNING YOUNG VEGETABLES
Although vegetables need space to develop, it is worth sowing salad crops more densely and thinning them when young. The thinnings, here of spring onions, can be used in salads while the rest of the crop matures.

Transplanting Self-sown Seedlings

Many plants, such as poppies (*Papaver*) or hellebores (*Helleborus*) naturally seed themselves about the garden. If you want the seedlings elsewhere, lift them, replanting them immediately into prepared soil in a suitable site. Keep watered and shaded, if necessary, until they are established.

LIFTING SEEDLINGS
Use a trowel to lift each seedling (here oriental poppy) with enough soil to avoid disturbing its roots. Replant to the same depth.

Transplanting Vegetable Seedlings

Although many vegetables are sown direct in drills (*see pp.54–55*), you may sow some in a seedbed to make the best use of space. (*For transplanting vegetables started in modules, see p.47.*)

Transplanting crops started elsewhere allows you to use a "sheet mulch" to control weeds around the crop. Cover the plot with a permeable membrane or biodegradable paper, and then transplant seedlings into the ground through slits in it.

MULCHED BED
Water the bed before laying the sheet: plant in the evening if you can, as the soil will stay moist and cool. Make a hole just large enough for the roots and set seedlings with the lowest leaves just above the sheet. Firm, label, and water in.

Transplanting Biennial Seedlings

Because they do not flower until their second year, biennials are often raised in nursery beds and transplanted to their final flowering positions when large enough. Seedlings from spring-sown seed, or self-sown seedlings, are planted out in a nursery bed in summer to grow on, and can then be placed in their final positions in the autumn.

1 **Sow biennials** (here wallflowers) in rows in a prepared seedbed. In a month or so, when the seedlings are 5–8cm (2–3in) tall, lift them with a fork.

2 **Plant out** the seedlings 15–20cm (6–8in) apart, in rows 20–30cm (8–12in) apart, in a nursery bed, allowing plenty of space for the roots. Firm in, label, and water.

3 **In autumn,** when the new plants are growing well, water the bed and carefully lift the plants. Transplant them into their flowering position in well-prepared soil.

PLANTS TO GROW FROM SEED

Plants that can be grown from seed range from annuals, biennials, perennials, bulbous plants, vegetables, and herbs, to trees, shrubs, and climbers. You can collect seed from your own garden, or buy them from a retail seed supplier.

☼ Prefers full sun ◐ Prefers partial shade ● Tolerates full shade ◊ Prefers well-drained soil ◊ Prefers moist soil ◆ Prefers wet soil ❂ Frost tender (min. 5°C/41°F) ❄ Half hardy (min. 0°C/32°F) ❄❄ Frost hardy (min. -5°C/23°F) ❄❄❄ Fully hardy (min. -15°C/5°F) ♈ RHS Award of Garden Merit

ANNUALS AND BIENNIALS

ALTHOUGH SHORT-LIVED, annual and biennial flowers are among the easiest plants to raise from seed and reward with a colourful, often long-lasting, display in spring or summer. Annuals flower and die within one year, while biennials take two years, flowering in their second year.

Ageratum houstonianum
(Floss flower)
Easily raised annual. Sow seeds from late winter to early spring. Raise under glass in cool and cold climates and provide a germination temperature of 20–25°C (68–77°F). Seeds take 10–14 days to germinate. Prick out seedlings within seven to ten days. Plants take at least 12 weeks to start flowering.
☼ ◊ ❄

Brachyscome iberidifolia
(Swan river daisy)
Moderately easy annual. Sow seeds from midwinter to early spring. Raise under glass in cool and cold climates. Surface sow because light is necessary for a good rate of germination, which takes from 7–21 days at a temperature of 18°C (64°F). Plants should flower 12–14 weeks after sowing.
☼ ◊ ❄

BRACTEANTHA BRACTEATA 'DARGAN HILL MONARCH'

Bracteantha bracteata
(Everlasting, helychrysum, strawflower)
Easily grown annual. Sow seeds in early to late spring, under glass in cool and cold climates. Cover seeds with their own depth of compost or vermiculite, as they need light. Germinates in seven days at a temperature of 15–21°C (59–70°F). Transplant within seven to ten days. Plants take 16–20 weeks to flower.
☼ ◊ ❄

◄ COLOURFUL HARVEST *A classic cornfield mixture with annual poppies and cornflowers*

Briza
(Quaking grass)
Sow the annual *B. maxima* and *B. minor* in situ in early autumn in areas with mild winters, or in mid-spring. Germination takes 10 days at 10°C (50°F). Seedlings usually flower within 14 weeks.
☼ ◊ ❄❄❄

***Calceolaria* Herbeohybrida Group**
(Pouch flower, slipper flower)
Biennials grown as flowering pot plants under glass. Sow in midsummer for flowers early next summer. Surface sow the very fine seeds as they need light to germinate. Germinates in 14–21 days at 15–21°C (59–70°F). Pot up seedlings in 7–10 days. Moderately easy.
◐ ♦ ❄

Calendula officinalis
(Pot marigold)
Sow in situ in autumn in mild climates, or in early to mid-spring to flower in 10–12 weeks. Germinates in seven days at 15–20°C (59–68°F). Self-sows freely, but seedlings do not come true to type.
☼ ◊ ❄❄❄

CALENDULA OFFICINALIS 'LEMON QUEEN'

Campanula medium
(Canterbury bells)
Easily raised biennial. Sow from late spring to early summer to flower the next summer. Surface sow the seeds because they need light to germinate. This takes 10–14 days under glass at 15–21°C (59–70°F). Transplant seedlings to a nursery bed. Plant in flowering positions in autumn.
◐ ◊–♦ ❄❄❄

Centaurea cyanus
(Cornflower)
Very easily raised annual. Sow in flowering position in early spring. May self-sow: seedlings come fairly true to type. Germination in 10–14 days at 10–15°C (50–59°F). Thin or transplant seedlings as necessary as soon as large enough to handle. Plants flower in 12 weeks.
☼ ◊ ❄❄❄

Consolida ajacis
(Larkspur)
Easily raised annual. Sow in situ during autumn in mild areas, or in early to late spring. Make successional sowings in spring. Seed sown at 10–15°C (50–59°F) takes 7–10 days to germinate. Autumn sowings flower in late spring, spring sowings in 12–16 weeks.
☼ ◊ ❄❄❄

Cosmos
Easily grown annuals. *C. bipinnatus* is the most popular; *C. sulphureus* is more tender. Sow under glass in mid-spring and plant out when danger of frost is over, or sow in situ in late spring. Germinate at 16°C (61°F). Flowers in 10–12 weeks.
☼ ◊ ❄ –⬘ (min. 5°C/41°F)

Dianthus
(Pink, sweet william)
Pinks, *D. chinensis*, are easily raised under glass in early spring to flower in summer. Germinate at 13–15°C (55–59°F). Sow biennial sweet william, *D. barbatus*, in early summer in an outdoor seed bed; transplant in autumn.
☼ ◊ ❄❄❄

Digitalis purpurea
(Foxglove)
Easily grown biennial. Sow in an outdoor seed bed in late spring. Germinates in seven days at 10–15°C (50–59°F). Transplant seedlings to nursery rows then to final positions in autumn. Self-sows readily. Flowers early summer.
◐ ◊–♦ ❄❄❄

Erysimum
(Wallflower)
Spring-flowering *E. cheiri* and *E.* x *allionii* 🏆 (Siberian wallflower) are easy biennials. Sow in a seed bed in late spring or early summer; germinates in 10–14 days at 10–15°C (50–59°F). Transplant to rows, then final positions in autumn.
☼ ◊ ❄❄❄

ERYSIMUM X *ALLIONII* 🏆

ESCHSCHOLZIA CALIFORNICA 🏆

Eschscholzia californica 🏆
(California poppy)
Easily grown annual. Sow in flowering position in early autumn or from early to late spring. In cool climates protect autumn sowings with cloches. Sow successive batches for a prolonged display. Germinates in 7–14 days at 10–15°C (50–59°F). Self sows freely. Spring sowings flower in 12–16 weeks.
☼ ◊ ❄❄❄

Felicia
(Blue daisy)
Includes easily grown annuals. Several species and cultivars are grown for summer flowers. In cool and cold climates sow seeds under glass from early to late spring. Germination takes up to 10 days at a temperature of 15–18°C (60–65°F). Plant out young plants when the danger of frost is over.
☼ ◊–♦ ❄ (min. 3–5°C/37–41°F)

Helianthus
(Sunflower)
Helianthus annuus and *H. debilis* are easily grown annuals. Sow in flowering

IMPATIENS WALLERIANA 'TEMPO LAVENDER' 🏆

positions in early spring, or sow under glass in late winter, singly in pots. Germination takes 7–10 days at an optimum temperature of 15°C (59°F). Sunflowers bloom in 16–20 weeks from sowing.
☼ ◊ ❄❄❄

Iberis
(Candytuft)
Iberis amara and *I. umbellata* are easily grown annuals. Sow seeds in flowering positions, during autumn in mild climates, or in succession from early spring to early summer. Seeds germinate readily in 7–10 days at a temperature of 10–15°C (50–59°F). Spring sowings take 12–16 weeks to flower.
☼ ◊ ❄❄❄

Impatiens
(Balsam, busy lizzie)
Impatiens balsamina (Indian balsam) and *I. walleriana* (busy lizzie) are easily grown, tender annuals. In cool and cold climates sow under glass from early to late spring. Germination requires a temperature of 15–18°C (60–65°F) and takes 7–10 days. Plant out when frosts are over. Plants take 12–16 weeks to reach flowering size.
☼–☀ ◊–♦ ❁ (min. 5°C/41°F)

Ipomoea
(Morning glory)
Ipomoea nil, I. purpurea and *I. tricolor* are easily grown, tender climbing annuals. In frost-prone climates sow under glass from mid-spring to early summer. Soak seeds in tepid water for 24 hours before sowing singly in pots and germinate at 18°C (64°F). This takes 7–10 days. Plants flower in 16 weeks.
☼ ◊ ❁ (min. 7°C/45°F)

Lathyrus odoratus
(Sweet pea)
Moderately easy annual. Sow in flowering positions in early to mid-spring, or autumn in mild areas. In cold regions sow in tube pots in autumn or winter and germinate in a cold frame. Optimum germination temperature is 10–15°C (50–59°F). Soak seeds overnight in tepid water. Germination takes 7–14 days.
☼ ♦ ❄❄❄

IPOMOEA TRICOLOR 'HEAVENLY BLUE' 🏆

Lunaria annua
(Honesty)
Easily raised biennial. Sow outdoors in early summer. Seed germinates in 7–14 days at a temperature of 10–15°C (50–59°F). If sown in a seedbed, transplant the seedlings within two weeks. Flowers in the following year, in late spring or early summer.
◐ ♦ ❄❄❄

Matthiola
(Gillyflower, stock)
Matthiola incana (gillyflower or Brompton stock) and *M. longipetala* subsp. *bicornis* (night-scented stock) are moderately easy annuals or biennials. Raise gillyflowers in pots under glass in spring. Sow night-scented stock in situ during spring. Seeds take 7–14 days to germinate at 10–15°C (50–59°F).
☼ ♦ ❄❄❄

Molucella laevis
(Bells of Ireland)
Easily raised annual. For best germination chill seeds at 1–5°C (34–41°F) for two weeks before sowing in early to mid-spring. In cool and cold climates raise plants under glass, germinating seeds at 13–18°C (55–64°F). Transplant as soon as large enough to handle easily. Flowering starts within about 12 weeks.
☼ ◊ ❄

MATTHIOLA INCANA
CINDERELLA SERIES ♀

NICOTIANA x *SANDERAE*
DOMINO SERIES

Myosotis sylvatica
(Forget-me-not)
Easily grown spring-flowering biennial. Sow in late spring or early summer in an outdoor seedbed. Germinates at 10–15°C (50–59°F) in about 14 days. Transplant seedlings to nursery rows and plant out young plants in autumn. Self-sows freely and comes reasonably true to type.
◐ ♦ ❄❄❄

Nicotiana
(Tobacco plant)
Nicotiana alata, *N.* x *sanderae*, and *N. sylvestris* ♀ are easy annuals. In frost-prone climates raise under glass in early to late spring. Seeds are fine and need light to germinate: mix with fine sand and surface sow. Germination takes seven days at 21°C (70°F). Plants flower in 12 weeks.
☼ ◊ ❄

Nigella damascena
(Love-in-a-mist)
Easily raised, hardy annual. Sow outdoors mid-autumn or spring: seeds need a soil temperature of 10–15°C (50–59°F) to germinate. If necessary, transplant within 14 days. Spring sowings flower in 16 weeks, autumn sowings the following spring.
☼ ◊ ❄❄❄

Papaver
(Poppy)
The biennial *P. croceum* (Arctic or Iceland poppy) and the annuals *P. rhoeas* (corn, field, or Flanders poppy) and *P. somniferum* ♀ (opium poppy) are all easily grown. Sow seed in flowering positions in spring (early summer for biennials). Seeds germinate in 7–14 days at a temperature of 10–15°C (50–59°F). Annuals flower in 12 weeks.
☼ ◊ ❄❄❄

Phlox drummondii
(Annual phlox)
Moderately easily raised, frost-hardy annual. In frost-prone climates sow the seeds

PHLOX DRUMMONDII
'LIGHT SALMON'

TROPAEOLUM 'PEACH MELBA'

under glass in early to late spring. Germination occurs within seven days at a temperature of 15–21°C (59–70°F). Prick out the seedlings within a week, and plant out the young plants when danger of frost has passed. Flowers appear in 12–16 weeks.

☼–◐ ◊–♦ ❄❄

Reseda odorata

(Mignonette)

Easily raised, hardy, fragrant annual. Sow seed in situ either in early to mid-autumn or in early to mid-spring. Germination occurs in 7–21 days at a temperature of 13–15°C (55–59°F). In areas prone to frost, protect autumn sowings with cloches or fleece. Plants from spring sowings flower in 12–16 weeks; plants from autumn sowings flower in the following spring.

☼ ◊ ❄❄❄

Rudbeckia hirta

(Black-eyed Susan)

Easily raised, hardy biennial with many cultivars, usually grown as annuals. Raise under glass in early spring, taking care not to sow seeds too deeply. Germination takes 7–14 days at a temperature of 6–18°C (61–64°F). Prick out the seedlings within seven days. Plants reach flowering size in 20 weeks.

☼ ◊ ❄❄❄

Salvia

(Sage, clary)

Salvia farinacea (mealy sage) and *S. splendens* (scarlet sage), which are half-hardy, and *S. viridis* (annual clary), which is fully hardy, are all easily raised annuals. Sow seed of half-hardy types under glass in spring; germination occurs in 7–21 days at a temperature of 10–15°C (50–59°F). Transplant the seedlings in 7–10 days. Sow seed of hardy annuals in their flowering positions during spring. Plants flower in 16 weeks.

☼ ◊ ❄ – ❄❄❄

Tagetes

(Marigold)

African (*T. erecta*), French (*T. patula*), Afro-French (*T. erecta* x *T. patula*) and Signet Group (*T. tenuifolia*) marigold cultivars are all easily raised, half-hardy annuals. In cool and cold climates sow seed under glass in spring and germinate at a temperature of 21°C (70°F). Germination time is seven days, and seedlings are vigorous; prick out within seven days. Flowers appear in 8–12 weeks.

☼ ◊ ❄

Tropaeolum

(Nasturtium, canary creeper)

The bushy or trailing nasturtiums derived from *T. majus* are easily grown half-hardy annuals. The canary creeper, *T. peregrinum*, is an easily raised tender annual. Sow seeds of either kind in early spring, or in final flowering position in mid-spring. Germination takes seven days at a temperature of at 13–16°C (55–61°F). Prick out seedlings within seven days, and plant out when the risk of frost is past. Plants reach flowering size in 12–16 weeks from sowing.

☼ ◊ ❄ (min. 3°C (37°F)

ZINNIA HAAGEANA 'ORANGE STAR

Zinnia

Moderately easy annuals, *Z. elegans* is tender and *Z. haageana* half-hardy. In frost-prone climates, sow under glass in early to mid-spring and germinate at 13–18°C (55–64°F). Germination occurs within seven days. Prick seedlings out into single pots or modules within seven days; zinnias dislike root disturbance, so pot seedlings singly into modules or biodegradable pots. Plant out when risk of frost is past. Zinnias flower in 16–20 weeks.

☼ ◊ ❄ –⊛ (min. 10°C/50°F)

VEGETABLES AND HERBS

THE MAJORITY OF VEGETABLES are raised from seed each year, bringing an edible harvest within a few months. Growing your own is easier than ever: modern cultivars are often resistant to pests or diseases, or less prone to bolting. To flavour your vegetables and other dishes, also raise some herbs from seed.

Aubergine
Grown as an annual for its summer fruits. Challenging: grow under glass in cool or cold climates, outdoors in warm areas. Sow in spring and germinate at 21°C (70°F). For the highest germination rates soak the seeds in warm water for 24 hours before sowing. Pot the seedlings as soon as they are large enough to handle.
☼ 💧 ⬙ (min. 25°C/77°F)

Beans
Easy annuals maturing in 13 weeks. Sow outdoors. Runner beans and French, kidney, or haricot beans are tender: sow these in mid-spring to early summer. They germinate at 12°C (54°F). Broad beans are hardy and can be sown in autumn or in spring, germinating at 10°C (50°F).
☼ 💧 ⬙ (min. 15°C/59°F)–❄❄❄

RUNNNER BEANS

Beetroot
Easily grown root crop with seeds that each produce several plants: rinse well in cold running water just before sowing for rapid germination. Sow outdoors. Needs at least 7°C (45°F) to germinate. Sow every three weeks from spring until midsummer. Harvest in 7–13 weeks.
☼ 💧 ❄❄❄

Carrot
Easily raised root crop. Sow in situ from spring to late summer. Soil must be above 7°C (45°F) for germination. Buy treated, primed seeds or fluid-sow for more even germination. Round-rooted types may also be multiblock sown under glass. Mature in 9–12 weeks.
☼ 💧 ❄❄❄

GROWING BRASSICAS FROM SEED

The brassica or cabbage family includes a wide range of biennial vegetables, including Brussels sprouts, cabbage, calabrese, cauliflower, kales, mustard, and sprouting broccoli, and the root vegetables swede and turnip. Some of these are grown as biennials for their shoots or flowerheads, while those grown for either their leaves or their roots are treated as annuals. Chinese cabbage withstands only light frosts and bolts above 25°C (77°F).

RED SUMMER CABBAGE

Most brassicas are cool-season crops of varying hardiness. All are easily raised, mainly by sowing seed in an outdoor seed bed in spring or summer and transplanting seedlings to their growing positions; they can also be raised in modules under glass. Swedes and turnips are sown in situ. Refer to seed packets for the precise times of sowing.

Crop rotation is vital for brassicas to avoid a build up of club root disease. If this is a problem, lime the soil in autumn and raise seedlings in modules to give plants a healthy start.
☼ 💧 ❄❄ – ❄❄❄

Celery
Self-blanching celery is a moderately easy stem vegetable, which can take light frosts. In cool regions celery is usually sown under glass in spring and planted out after frosts. Seeds need light and a minimum of 15°C (59°F) to germinate. Raise in modules in order to prevent root disturbance.
☼ 💧 ❄

Chard
A leafy "cut-and-come-again" vegetable with white or coloured stems. Biennial, hardy (a cool-season crop), and easily grown, seeds are sown in situ and can be sown in mid- to late spring for summer and autumn cropping, or in early autumn to give a spring crop. The seeds need a temperature of at least 7°C (45°F) to germinate.
☼ 💧 ❄❄❄

Cress
This is usually grown as a seedling crop together with mustard. An easy, moderately hardy annual. Grow under glass or indoors, sowing on moist, absorbent paper at any time of year, and germinate at 15–20°C (59–68°F). Ready in ten days. Seeds can also be broadcast sown outdoors during spring, late summer, or autumn.
☼–☀ 💧 ❄❄

CUCUMBERS

LAND CRESS

Cucumber
Moderately easy, tender, annual climber grown for its fruits. Grows best at 20–30°C (70–89°F). Sow seeds in spring, one per small pot. Needs a temperature of 20°C (70°F) to germinate. Plant out when large enough, in cool climates either under glass or outdoors when frost has passed, depending on the cultivar. Start harvesting within 12 weeks.
☼ 💧 ❄ – (min. 20°C/70°F)

Lettuce
Easy salad vegetable with a range to suit different seasons of sowing and harvesting, allowing for year-round cropping. In frosty climates seeds can be grown under glass in winter, outdoors at other times. Seeds do not germinate above 25°C (77°F). Sow in situ or raise in modules. Main sowing period early spring to early autumn.
☼ 💧 ❄❄ – ❄❄❄

Land cress
This is an easy salad vegetable similar to watercress, and grown as an annual or biennial. In cool and cold climates sow seed at 10–15°C (50–59°F) in mid- or late summer for autumn to spring crops, or in mid-spring to early summer for autumn crops (tends to bolt). Sow in situ or in seed trays under cover. Pick leaves from seven weeks on.
☼–☀ 💧 ❄❄❄

Onions and leeks
Hardy, easily grown, cool-season annuals or biennials. Sow bulb and spring onions in early spring. Sow bulb onions either direct or in modules or multiblocks under glass. Sow leeks in an outdoor seed bed or in modules in early spring and transplant into deep holes. Germination temperature for all seeds is 10–15°C (50–59°F).
☼ 💧 ❄❄❄

LEEKS

Parsnip
Moderately easy, cool-season root crop. Use fresh seeds to obtain the best germination rates, and pregerminated or primed seeds for more even germination. Sow in situ in early or late spring. Seeds need at least 12°C (54°F) to germinate. Harvest roots from 16 weeks.
☼ ◊ ❄❄❄

Peas
Podded vegetables including mangetout and sugar peas. Moderately easy, cool-season annuals. Sow seeds in situ from spring to early summer, or in autumn for overwintering crops of very hardy cultivars. Successional spring and summer sowings can be made every ten days. A soil temperature of 10°C (50°F) is needed for germination. Start to harvest spring or summer sowings in 10–12 weeks.
☼ ◊ ❄ – ❄❄

Peppers
Annual fruiting vegetables including both sweet or bell peppers and chilli peppers. Moderately easy to raise.

PEAS

SWEET PEPPERS

Often grown under glass in cool climates. Sow in pots, in early spring for greenhouse crops or in mid-spring for outdoor crops. Seeds need a temperature of 21°C (70°F) to germinate. Pot off seedlings at four-leaf stage.
☼ ◊ ❖ (min. 21°C/70°F)

Pumpkins and winter squashes
Easily grown, annual fruiting vegetables. In frost-prone climates raise in pots under glass in early spring and plant out when frosts are over. Soak seeds overnight in water before sowing and sow three seeds to a small pot. Seeds need a temperature of 21°C (70°F) to germinate. Thin to leave the strongest seedling.
☼ ◊ ❖ (min. 18°C/64°F)

Radish
Summer, winter, and oriental (mooli) radishes are easy annual root crops. The winter and oriental types are frost hardy. Sow seed in situ, from spring to late summer for summer radishes, and in summer for winter and oriental radishes. Sow seeds of summer radish in succession every 10 days. Summer radishes are ready in three to four weeks, while the oriental types take eight weeks. Winter radishes take 12 weeks.
☼ ◊ ❄ – ❄❄

Spinach
An easy, annual, leafy, cool-season crop. Seeds of spinach are difficult to germinate above 30°C (86°F). Sow seed in situ at three-weekly intervals from late winter through to midsummer, using bolt-resistant summer cultivars. Begin harvesting the leaves in six to eight weeks. The hardier cultivars can also be sown in situ during early autumn to allow for spring harvesting.
☼ – ◐ ◊ ❄❄ – ❄❄❄

Summer squashes, courgettes, and marrows
Easily grown, annual fruiting vegetables. In frost-prone climates raise plants in pots under glass in early spring and plant out when frosts are over. Sow three seeds to a small pot and germinate at 15°C (59°F), thinning to leave

MARROW

the strongest seedling. Seeds can be sown outdoors in situ in late spring.
(min. 18°C/64°F)

Sweet corn
Easily raised annual grown for its cobs. Seeds germinates at a temperature of 10°C (50°F). Sow in spring. In cool climates sow singly in modules under glass and plant out seedlings within two weeks, protecting with cloches to start with, if necessary. Plant in blocks rather than rows for good pollination. Cobs are produced in summer.
(min. 16°C/61°F).

Tomato
Easy fruiting vegetable grown as an annual for its summer fruits. In cool climates grow under glass, or outdoors in summer using suitable cultivars. Sow in early to mid-spring in modules. Seeds germinate at a temperature of 15°C (59°F). Pot seedlings into small pots to grow on before planting out. Tomatoes are ready to harvest from seven to eight weeks onwards.
(min. 21°C/70°F)

TOMATO

HERBS FROM SEED

BORAGE

Basil
Easy annual. Sow under cover at a temperature of 18°C (64°F) in late spring, or outdoors at 15°C (59°F) in early summer.

Borage
Easy annual. Sow outdoors in early to late spring. Tap-rooted, so avoid disturbance. Self-seeds, and can become a weed.

Coriander
Moderately easy annual. Sow in situ in early or late spring. Needs full sun for seed production. May self-seed in suitable conditions.

Fennel
Easy perennial. Sow in early spring in pots with bottom heat of 15–21°C (59–70°F). Plant out in late spring. Tends to self-seed around the garden.

Oregano or marjoram
Easy perennial. Surface sow thinly in spring under glass. Germinates at a temperature of 10–13°C (50–55°F), but is often erratic.

Parsley
Easy annual. Sow in early spring with bottom heat of 18°C (64°F), or in late spring in situ at 15°C (59°F). Germination is slow.

Sage
Easy perennial. Sow in modules under cover in early spring, with bottom heat of 15°C (59°F), and cover with perlite. Plant out in late spring.

Thyme
Easy perennial. Surface sow seeds in containers in spring with bottom heat of 20°C (68°F), or in situ in late spring or early summer.

PARSLEY

PERENNIALS AND BULBOUS PLANTS

SEEDS PROVIDE AN ECONOMICAL way of raising large numbers of perennials, but bear in mind that many hybrids and cultivars will not come true to type. Seeds also enable gardeners to replace short-lived perennials as they deteriorate, and offer the best means of increasing slow-growing plants.

Achillea
(Yarrow)
Easily raised, with both border and alpine types. Sow seeds in autumn or spring in pots under glass, and germinate them at a temperature of 15°C (59°F). Pot up seedlings as soon as possible and plant out when large enough. Seedlings often flower in their first year, especially border types.
☼ ◊ ❄❄❄

Alchemilla
(Lady's mantle)
These mainly hardy perennials, such as the popular *A. mollis* 🏆, are easy to raise. With hardy species best results are obtained from an autumn sowing followed by exposure to winter cold. Most will also readily self-sow. Sow seed of frost-hardy species such as *A. ellenbeckii* in spring and germinate at 15°C (59°F).
☼ ♦ ❄ – ❄❄❄

Anemone
(Windflower)
The herbaceous species such as *A. hupehensis* are moderately easy to raise from seed. For best germination sow seeds in pots as soon as ripe and place in a cold frame. Fresh, spring-sown seeds should be germinated at 15°C (59°F). Expect flowering in second or third season.
☼–◐ ♦ ❄❄❄

Antirrhinum majus
(Snapdragon)
Short lived, easily raised perennials, often grown as summer bedding plants. In frost-prone climates sow seeds under glass in early spring. Germinate at 16–18°C (61–64°F). Prick out into trays or modules, and plant out when danger of frost is over. Flowers in summer.
☼ ◊ ❄

Aquilegia
(Columbine)
Easily raised. Sow fresh seeds in late spring or early summer and germinate at 10°C (50°F). Sow stored seeds in autumn and expose to winter cold. Aquilegias hybridize and self-sow very freely. Collect seeds only from isolated plants. Flowers in second or third year.
☼–◐ ♦ ❄❄❄

ANEMONE HUPEHENSIS

AQUILEGIA MCKANA GROUP

Athyrium filix-femina 🏆
(Lady fern)
Deciduous fern. Sow spores as soon as ripe under glass and germinate at a temperature of 15°C (59°F). This is one of the easier ferns to raise, and young plants are produced surprisingly quickly. (*For full details of raising ferns from spores see pp.48–49.*)
◐ ♦ ❄❄❄

Aubrieta
(Aubretia)
Easily raised rock-garden perennials. Only cultivars are grown, and are best raised from packet seed; seedlings from home-saved seed will vary. Sow in pots when ripe or in early spring. Germinate in a cold frame. Prick out seedlings into small pots, and plant out when large enough. May flower in second season.
☼ ◊ ❄❄❄

Begonia semperflorens
Moderately easy tender perennial, with cultivars generally grown for summer bedding. In frost-prone climates sow seed under glass in early spring. The very fine seed needs light to germinate, so do not cover. Germination at 21°C (70°F), takes two to three weeks. Prick out into trays or modules and plant out after frosts. Flowers appear throughout summer.
◐ ◊ ❅ (min. 5°C (41°F)

Bellis perennis
(Daisy)
Easily raised perennial, grown as a biennial for spring bedding from packet seed. Sow in an outdoor seed bed in early summer. Seeds need 10°C (50°F) to germinate. Transplant seedlings to nursery rows and move young plants to flowering positions in autumn. Flowers produced the following spring.
☼–◐ ◊ ❄❄❄

Campanula
(Bellflower)
Herbaceous plants and alpines, easily raised from seed. Some self sow freely. Sow the fine seeds thinly and cover lightly. Keep spring-sown seeds at 15°C (59°F) and autumn-sown alpines such as *C. carpatica* 🏆 in a cold frame. Prick out into individual pots to grow on and plant out when sufficiently large. Some may flower in second year.
☼–◐ ◊ ❄❄❄

Euphorbia
(Spurge)
Easily raised perennials. Sow seeds at 15°C (59°F) in spring. Germination can be erratic;

EUPHORBIA SIKKIMENSIS

seedlings may appear over several months. To overcome this, sow fresh seed in autumn and expose to winter cold. Seeds should then germinate more evenly in spring. Plants take two to three years to reach flowering size.
☼–◐ ◊–💧 ❄❄❄

Festuca
(Fescue)
Easily raised ornamental grasses. Sow seeds in pots in a cold frame during autumn, or in spring outdoors. Minimum temperature of 10°C (50°F) needed to germinate; fresh seed germinates rapidly. If raised in pots, pot the seedlings singly to grow on. Makes substantial clumps in two to three years.
☼ ◊ ❄❄❄

Gazania
Easily raised evergreen perennials, usually grown as annuals for summer bedding. Mainly cultivars are grown – buy packet seed. In frost-prone climates sow under glass in late winter or early spring. Germinate at 18–20°C (64–68°F). Pot up seedlings singly and plant out after frosts. Flowers are produced from early summer.
☼ ◊ ❄

Geranium
(Cranesbill)
Easily raised perennials. Sow seeds as soon as ripe or in early spring and germinate in a cold frame. Germination takes 14 days at 15°C (59°F) . Species hybridize readily, and some self-sow. Some cultivars, such as *C. wallichianum* 'Buxton's Variety' 🏆, come fairly true from seed. Many flower the following year.
☼–◐ ◊ ❄❄❄

Helleborus
(Hellebore)
Easily raised from seed. Some self sow freely. *H. hybridus* hybridizes freely, but hybrid seedlings are usually attractive. Sow seeds in pots as soon as ripe and expose to winter cold to break dormancy. They may germinate in autumn or wait until spring. Flowers are produced in two to three years.
◐ 💧 ❄❄ –❄❄❄

GAZANIA 'TALENT YELLOW'

Kniphofia
(Red hot poker)
Easily raised perennials. Sow seeds in containers under glass in spring. Germinate at 18°C (64°F). Pot up seedlings singly to grow on. Hybrids seldom come true but can be bought as packet seed: some flower in first year if sown early. Species flower in two to three years.
☼ ◊ ❄❄ – ❄❄❄

Mammillaria
(Pincushion cactus)
Easily raised clump-forming cacti. Self-fertile species often set seed; collect when pods are soft or buy packet seed. Sow fresh seed under glass from spring to autumn. Germinate at 19–24°C (66–75°F). Prick out into very small pots when large enough to handle. Flowers in two to five years.
☼ ◊ ❂ (min. 7-10°C/45-50°F)

Paeonia
(Peony)
Herbaceous peonies are moderately easy to raise. Seeds are doubly dormant; chill for several weeks before sowing or sow in pots as soon as ripe and leave outdoors in the winter cold. Seeds need a second period of cold before shoots appear. Five years to reach flowering size.
☼ ◊ ❄❄ – ❄❄❄

MAMMILLARIA MAGNIMAMMA

PRIMULA 'DREAMER'

Papaver orientale
(Oriental poppy)
These poppies seed freely and are easily raised. Sow in spring or summer as soon as ripe. Seeds need light to germinate, so surface sow. Germination in 10 days at 50°F (10°C). Transplant as soon as large enough to handle – poppies dislike root disturbance. Many flower in following season.
☼ ◊ ❄❄❄

Pelargonium
("Bedding geranium")
Seed-raised pelargoniums are easily raised and widely treated as annuals for summer bedding. In frost-prone climates sow under glass in late winter and germinate at 21°C (70°F). This takes 7–10 days. Pot seedlings to grow on under glass and plant out when danger of frost is over.
☼ ◊ ❂ (min. 2°C (36°F)

Petunia
Petunias are popular summer bedding plants, generally grown as annuals. Easily raised. In frost-prone climates sow under glass in mid-spring and germinate at 15°C (59°F). This takes ten days. Prick out into trays or modules and plant out when frosts are over to flower in the summer.
☼ ◊ ❄

Primula
The variable herbaceous perennials and alpines are easy from fresh seed (seeds are short-lived). Many species hybridize freely. Sow seeds in pots as soon as ripe, or in spring. They need light, so cover them only lightly with vermiculite, and germinate in an open cold frame. Some will flower the following season.
◐ ♦ ❄❄❄

Thalictrum
(Meadow rue)
Moderately easy to raise. Sow seeds as soon as ripe, or in early spring. Older seed germinates erratically. Sow in pots or pans and germinate in

THALICTRUM KIUSIANUM

VERBENA × HYBRIDA
ROMANCE SERIES

a cold frame. Pot seedlings to grow on. Plants take two to three years to flower.

Verbena
(Vervain)
The verbenas grown for summer bedding are usually treated as annuals. Easily raised. In frost-prone climates sow seeds under glass in early spring. Germination takes 14 days at 21°C (70°F). Prick out into pots or trays and plant out when frosts are over. Flowers from early summer.

Viola
(Pansy, violet)
Easy, but may be short-lived. Sow most species in early to mid-spring, in pots or trays and germinate at 15°C (59°F). Sow winter-flowering pansies in mid-summer. Germination in 10–14 days. Transplant when large enough to handle. Some species self-sow and hybridize freely. Most flower in following or same season.

BULBOUS PLANTS FROM SEED

Crocus
Easy cormous perennials. Sow seed in trays in late summer and germinate in a cold frame. Corms can be planted out after two years, and reach flowering size after three years.

Cyclamen
(Sowbread)
Easy tuberous perennials. Sow seeds of hardy cyclamen as soon as ripe, after soaking in warm water for 12 hours. Germinate in a cold frame. Transplant the seedlings as soon as they are large enough to handle

Dahlia
Easy tuberous perennials. Seed-raised bedding dahlias are grown as annuals. Sow under glass in spring. Germinate at 16°C (61°F). Pot up and plant out after frosts are over.

LILUM SUPERBUM

NARCISSUS CYCLAMINEUS

Galanthus
(Snowdrop)
Easy bulbous plants. Species hybridize freely. Sow seed in pots as soon as ripe and germinate in open cold frame. Pot on after one year; bulbs reach flowering size after three years.

Lilium
(Lily)
Mainly hardy bulbs, easily raised from seed sown as soon as ripe. Germinate in cold frame. Many germinate following spring. Pot on seedling bulbs regularly. Flower in four to five years.

Narcissus
(Daffodil)
Easy bulbous plants. Species self sow readily. Sow seed in deep pots as soon as ripe and germinate in open cold frame. Pot on after one year. Bulbs reach flowering size in two to four years.

TREES, SHRUBS, AND CLIMBERS

TREES, SHRUBS, AND WOODY CLIMBERS form the backbone of any garden planting. They can be expensive to buy but generally are not difficult to propagate. Bear in mind, though, that they can be slow growing, particularly some trees, which may take many years to flower.

Acer
(Maple)
Easily raised trees. Sow ripe seeds in mid- to late autumn in a seed bed or pots in a cold frame, or store in moist peat in a refrigerator and sow in spring. Germinates at 10–15°C (50–59°F) but often not until the second year.
☼–◐ ◊ ❄❄ – ❄❄❄

Berberis
(Barberry)
Easy shrubs. Chill seeds from ripe fruits over winter: layer berries in sand in containers, or sow in pots, and keep in a cold place outside to germinate by summer. Sow stratified seed in an outdoor bed in spring.
☼ ◊ ❄❄ – ❄❄❄

Buddleja
(Butterfly bush)
Hardy species are easy. Seeds of *B. davidii* hybrids are often available. Sow in spring in a seed bed when the soil reaches 10°C (50°F). Plant in final position in autumn. *B. davidii* hybrids may flower in their second year.
☼ ◊ ❄❄ – ❄❄❄

Camellia
Easily raised shrubs. Sow ripe seed in autumn or spring, after soaking in hot water to soften the coats. Germinate under glass at 15–18°C (59–65°F). Pot singly as soon as large enough to handle.
◐ ◊ ❂ – ❄❄❄

BERBERIS WILSONIAE

Chamaecyparis
(Cypress)
Easily raised conifers; only raise species from seed. Extract seeds in autumn from one-year-old, ripening female cones, dry thoroughly, and store in the refrigerator at 1–4°C (34–39°F). Sow in spring with bottom heat of about 15°C (59°F). Transplant seedlings in midsummer.
☼ ◊ ❄❄❄

Citrus
(Lemons, oranges, grapefruit)
Moderately easy trees. A citrus seed produces several seedlings, some of which are clones of the parent. Sow in pots in summer and germinate at 16°C (61°F). Weed out puny or very vigorous seedlings to leave clones. Flowers in seven years.
☼ ◊ ❂ (min. 3–5°C/37–41°F)

Cotoneaster
Easily raised shrubs, although not generally true to type. In autumn and winter first give seeds 4–12 weeks at 20–25°C (68–77°F), then cold stratify. Sow in a seed bed in spring to germinate in the following year.
☼ ◊ ❄❄❄

CLIMBING PLANTS FROM SEED

Clematis
Sow in shallow pots in autumn as soon as ripe. Cover lightly with compost, topdress with grit, and place in a cold spot outdoors. Pot when large enough to handle.
☼ ◊ ❄ – ❄❄❄

Eccremocarpus scaber ♀
(Chilean glory flower)
Sow under glass in late winter or early spring and germinate at 13–16°C (55–61°F). Cover lightly. Pot as soon as large enough to handle. Plant out late spring.
☼ ◊ ❄❄

Passiflora
(Passion flower)
Store ripe fruit for 14 days, mash and leave in warm spot for three days. Clean seeds in running water and dry. Soak for 24 hours in hot water and sow at 20–25°C (68–77°F).

Cytisus
(Broom)
Easily raised shrubs. Sow seeds in pots outdoors as soon as ripe to germinate in spring. Pot up seedlings into small pots and plant out in autumn. Soak spring-sown seeds in hot water before sowing to soften the seed coats. Cytisus usually flower within two years.
☼ 💧 ❄ – ❄❄❄

Daphne
Easily raised shrubs: *D. mezereum* is often seed raised. Sow seeds as soon as ripe in summer or autumn, in pots of gritty seed compost, and place in a cool, frost-free place. Most germinate in spring after a winter's chilling. Leave for another year to germinate all the seeds.
☼–◑ 💧 ❄❄ – ❄❄❄

Eucalyptus
(Gum)
Fast-growing trees easily raised from seed. Seeds need a period at 3–5°C (37–41°F) for two months before sowing in spring. They dislike root disturbance, so sow in root trainers, a pinch of seed in each cell. Germination is quick at 15–20°C (59–68°F). Plant out in 12–15 months.
☼ 💧 ❄ (min. 5°C/41°F)– ❄❄❄

Grevillea
Shrubs that are moderately easy from seed. Sow as soon as ripe, or soak older seed in tepid water overnight before sowing. Use lime-free compost. Germinate at 15°C (59°F). Seed often germinates erratically, some species taking longer than others. Only *G. robusta* germinates easily.
☼ 💧 ❄ (min. 5°C/41°F)– ❄❄

Ilex
(Holly)
Mainly hardy trees whose seeds germinate readily, if slowly – up to three years. Collect berries in winter, clean off the flesh and sow at once in a seed bed. Alternatively store seeds in a warm, moist place, then chill them in moist compost in the refrigerator before sowing in spring.
☼–◑ 💧 ❄❄ – ❄❄❄

Leptospermum
(Tea tree)
Hardy to tender shrubs that are easily raised from seed. Sow in autumn or spring, under glass in cool and cold climates, and germinate in a temperature of 13–16°C (55–61°F). Pot off seedlings as soon as possible and plant out when large enough. Flowers from two to three years.
☼ 💧 ❄ (min. 5°C/41°F)– ❄❄

Prunus
(Cherries and plums)
Trees and shrubs easily raised from seed. Chill cleaned, moistened seeds in a sealed plastic bag in a refrigerator for six weeks at 1–4°C (34–39°F) before sowing in a seed bed in late winter. Transplant or pot up the seedlings as soon as possible.
☼ 💧 ❄❄ – ❄❄❄

GREVILLEA JUNIPERINA

PRUNUS SARGENTII

Rhododendron
Mainly hardy shrubs, easily raised from seed under glass in winter or early spring. Surface sow the fine seeds onto sieved, lime-free compost. To prevent drying out, place containers under mist, glass, or plastic film. Germinate at 16°C (61°F). First flowering from three to five years or more, depending on species.
◑ 💧–💧 ❄❄ – ❄❄❄

Sorbus
(Mountain ash)
Trees, easily raised from seed. Collect seed in autumn before berries ripen and germination inhibitors develop, and sow immediately in an outdoor seed bed. Alternatively chill seed for two months at 1–4°C (34–39°F) in a refrigerator (place seeds on moist blotting paper in a saucer) before sowing in late winter. Transplant in late spring.
☼–◑ 💧 ❄❄ – ❄❄❄

INDEX

Page numbers in ***bold italics*** refer to illustrated information

ACKNOWLEDGMENTS

Picture Research Brenda Clynch
DK Picture Library Romaine Werblow
Index Hilary Bird

Dorling Kindersley would like to thank:
All staff at the RHS, in particular Barbara Haynes and Susanne Mitchell.

The Royal Horticultural Society
To learn more about the work of the Society, visit the RHS on the internet at **www.rhs.org.uk**. Information includes news of events around the country, a horticultural database, international plant registers, results of plant trials and membership details.

Photography
The publisher would like to thank the following for their kind permission to reproduce their photographs:

(key: t=top, c=centre, b=bottom, l=left, r=right, fj=front jacket, bj=back jacket)

AKG London: Erich Lessing 9cr.
Bruce Coleman Ltd: Hans Reinhard 62.
Garden Picture Library: David Askham 9bl; Vaughan Fleming 10; A. I. Lord 13br; Steven Wooster 6.
John Glover: 5bl, 12, 14, 15br, 17t.
Jacqui Hurst: bj tr; 25.
N.H.P.A.: Martin Harvey 8; 21.
N.H.P.A.: Eric Soder fj bl.
Clive Nichols: fj r; Cecily Hill House, Gloucestershire 4br, 13tl; Copton Ash, Kent 15tl.
Oxford Scientific Films: John McCammon 22.